Advanced Traveler Information Systems

For a listing of recent titles in the *Artech House ITS Library*, turn to the back of this book.

Advanced Traveler Information Systems

Bob McQueen
Rick Schuman
Kan Chen

Artech House
Boston • London
www.artechhouse.com

Library of Congress Cataloging-in-Publication Data
McQueen, Bob.
    Advanced traveler information systems / Bob McQueen, Rick Schuman, Kan Chen.
        p.   cm. — (Artech House ITS library)
    Includes bibliographical references and index.
    ISBN 1-58053-133-4 (alk. paper)
        1. Route choice—Data processing.   2. Transportation—Data processing.
    3. Wireless communication systems.   I. Schuman, Rick.   II. Chen, Kan, 1928–.
    III. Title.   IV. Series.
    HE336.R68 M35   2002
    338.3'124—dc21   2002074694

British Library Cataloguing in Publication Data
McQueen, Bob
    Advanced traveler information systems. — (Artech House ITS
    library)
    1. Automatic driving—Electronic information resources
    2. Intelligent Vehicle Highway Systems   3. Automobiles—
    Electronic equipment
    I. Title   II. Schuman, Rick   III. Chen, Kan
    388.3'124

    ISBN 1-58053-133-4

Cover design by Yekaterina Ratner

© 2002 ARTECH HOUSE, INC.
685 Canton Street
Norwood, MA 02062

International Standard Book Number: 1-58053-133-4
Library of Congress Catalog Card Number: 2002074694

10 9 8 7 6 5 4 3 2 1

Contents

Acknowledgments

There have been so many influences on the development of the material in this book that it would be impossible to list all the people involved. Here are a few of the many.

Carl Ahlert	Steve Albert	Roger Allport
Gad Bachrach	Philip Banks	Richard Barber
Bob Barrett	Rena Barta	Chris Bausher
Susan Beaty	Wei Bin Zhang	Dick Bishop
Tom Black	Jim Blain	Larry Boatman
Adam Brand	Anne Brewer	Arnie Bloch
Jacov Blum	Richard Bossom	Serge Bueno
Chester Chandler	Mel Cheslow	Jon Cheyne
Belle Cole	John Collura	John Cox
Bud Cribbs	Brad Dennard	Khaled Dessouky
John Earp	Bob Easter	Bruce Eisenhart
Don Erwin	Gary Euler	Fred Ferrell
Jorge Figueredo	Gordon Fink	Elizabeth Fischer
Stig Franzen	Eloise Freeman	Jonathan Gifford
Ron Giguere	George Gilhooley	Bob Glass
Gene Glotzbach	L.A. Griffith	Mohammed Hadi

Abdul Rahman Hamad	Sharon Hansen	Greg Hatcher
Glenn Havinoviski	Ron Heft	Cliff Heiss
Martin Herbert	Tom Horan	Tom Humphrey
Carol Jacoby	Rob Jaffe	Keith Jasper
Ed Jensen	David Jeffrey	Allan Jepps
Christine Johnson	William Johnson	Bill Jones
Job Klinjhout	Ron Knockheart	Mike Krueger
Tom Lambert	Steven Landau	Jane Lappin
Barney Legge	Eva Lerner-Lam	Mac Lister
Steve Lockwood	Shelley Row	Joel Markowitz
Vito Mauro	John McGowan	Mike McGurrin
John Miles	Khairun Zainal Mokhtar	Bob Parsons
Bob Paulsen	Scott Perley	Mike Pietrzyk
Jerry Pittenger	Vic Poteat	Pierre Pretorious
Frik Rheeder	Marlin Ristenbatt	Craig Roberts
Bob Rupert	Siegfried Rupprecht	Eric Sampson
Don Savitt	Lyle Saxton	George Schoene
Joe Schuerger	Mike Shagrin	Benny Shalita
Terry Shaw	Eli Sherer	Steven Shladover
Lee Simmons	Loyd Smith	Ted Smith
Joe Schuerger	Jerry Sobetski	Heinz Sodeikat
Bill Spreitzer	Henk Staats	Ray Starsman
Henk Stoelhorst	Gloria Stoppenhagen	Larry Sweeney
Paul Taylor	Ilse Van Goth	Charlie Wallace
Chip White	Ladon White	Paul White
Malcolm Williams	John Wootton	Hal Worrall
Jim Wright	Karl Wunderlich	Oliver Yandle
Carol Zimmerman		

We acknowledge all of the other people we have learned from and been influenced by over the course of developing this book, with a special mention to all the participants in the ITS America Advanced Traveler Information Systems Committee and the Payment Systems Task Force. An extra special mention also for Judy McQueen for her painstaking and thorough review of the manuscript and for being Bob's best friend and partner.

Authors' contact information:

Bob McQueen—PBS&J, 407-647-7275 x328, e-mail bobmcqueen@pbsj.com

Rick Schuman—PBS&J, 407-647-7275 x511, e-mail rickschuman@pbsj.com

Kan Chen—650-375-8890, e-mail kanchen@attbi.net

1

Introduction

1.1 Learning Objectives

After you have read this chapter you should be able to do the following:

- Understand the background to the development of the book;
- Understand the structure and content of subsequent chapters;
- Define the target readers of this book;
- Explain why each target reader group will find it useful;
- Explain our proposed vision for the future of advanced traveler information systems (ATIS).

1.2 Background

ATIS make use of information and communication technologies to deliver information to a wide range of travelers who use different modes of travel and have a wide variety of characteristics. Information enables travelers to make better travel choices and supports better use of existing transportation facilities. Such systems, therefore, have the potential to improve the operation of the transportation network and raise the perceived quality of the traveler's experience through the provision of good usage and option information. When planned, developed, and implemented in a thoughtful, well-structured manner, ATIS can present private-sector business opportunities

1

while simultaneously satisfying a range of predefined transportation policy objectives. The primary objective of this book is to define, explain, and promote a coherent, orderly approach to the understanding of ATIS that can support the attainment of both private- and public-sector objectives. In doing this, we hope to encourage the adoption of approaches to the development and implementation of traveler information systems that address all modes of travel and a broad range of traveler types, while taking best advantage of private-sector motivations and interests.

The primary focus for our interest is the open ATIS that requires inter-organizational collaboration and coordination in order to operate. Such systems span jurisdictions and organizational boundaries. Closed ATISs are entirely operated and managed within a single entity solely for the purposes of that entity. While this latter category represents a significant segment of the market, there are a greater number of issues and challenges associated with the former.

When we first began discussing the writing of this book, we quickly developed a common view of what we wanted to achieve; that is, to deliver an informative and insightful source of knowledge, lessons, and experience drawn from our collective involvement in traveler information systems and intelligent transportation systems (ITS) generally over the past 15 years. We also aim to ensure that the material contained in the book is as accessible as possible, in terms of the use of straightforward language, writing in an informal, friendly, easy-to-read style, and structuring the materials to enable concise navigation through the flow of ideas and concepts. ITS and traveler information systems seem to be surrounded by a sea of complex terminology, which makes the explanation and exploration of the underlying concepts and philosophies difficult. In many cases, we bring this difficulty and complexity upon ourselves by making use of shorthand and jargon, which is very effective and efficient when used as code between professionals with a strong current knowledge and awareness of the meaning and appropriate usage of the code.

The motivation or desire to develop the materials for this book arose from our collective experiences in developing, planning, and applying information and communication solutions for transportation problems. In the early days of this experience, we all shared a view that one of the most significant and potentially beneficial applications of ITS would be the so-called ATIS. In the beginning we believed that this would come through the development and implementation of specific in-vehicle information systems that were custom developed for the delivery of traveler information to the driver in the vehicle. Static or historically derived route guidance and yellow pages information was assumed to be the first wave, to be rapidly superseded by dynamic route guidance and real-time in-vehicle information services. Dynamic route guidance would be linked to traffic-signal control, transit, and freeway-management systems to deliver to the driver information on the right route and modal choice for the prevailing transportation conditions. It was assumed that all these services and the supporting infrastructure would be developed and installed specifically for traveler information purposes.

Then along came the Internet and the World Wide Web (WWW, or the Web). Our paradigm of ITS-specific systems was blown apart as we realized that the information we sought was actually going to be delivered alongside many other data and information streams as part of this public access, multiple-application, global communications network. At first we understood that home and office delivery of the traveler information services we wanted would be delivered by information service providers peddling a portfolio of stock quotes, sports scores, weather, and traveler information to the Internet masses. Later, we realized that the Internet would become much more ubiquitous with the advent of mobile and handheld wireless Internet access. Then we thought we had seen the answer—in-vehicle mobile Internet access for many, many services, carrying traveler information to the driver in the driving seat as well as the traveler at home, in the office, and on the move in the course of the journey. We witnessed the embryonic beginnings of the traveler information service provider's business thinking: This is it, our dreams are about to come true, this is the start of a mega market.

Then, a strange thing happened, or rather did not happen. The traveler information services market did not climb to maturity as we had planned. Our infant market stalled, stunted by some basic lack of fuel or nutrients, sputtering along below critical mass. From the public-sector perspective, the ability to address transportation policy objectives and influence the behavior of the traveler through the availability of accurate and relevant information on choices and options became a lonely task with little external or private-sector involvement. From the private-sector perspective, certain core data that would be provided by the public sector either did not materialize or proved to be too expensive or time-consuming to mine and exploit. One of the centerpieces of our future big-picture plans for advancing the quality of transportation operations and management seemed as far from realization as ever.

So what are we really trying to do by pulling these materials together and developing the structure of this book? When we asked ourselves this question, we tried to get beyond the initial thought to identify and define the real value that we wanted to provide.

Being deeply involved in the market as practitioners and advisors to both public and private sectors, we became acutely interested in studying the traveler information issues, trying to find the root causes and the barriers to market maturity, trying to understand the mechanisms that make or break the market for traveler information. Working with major players on both sides, implementing projects, scanning for lessons learned and practical experiences, we slowly built what we think is a good and useful picture of what the market is, who the players are, what constitutes an ATIS, as well as some insight into the most effective ways of going about it from both public and private perspectives. We felt that it would be useful to share this information with a wider group of people than we usually come into contact with, hence the motivation to write this book. We may be eternal optimists, but we think that the traveler information market is alive and well and that the opportunity and promise we all perceived in the beginning is still there and still attainable. We believe that there are unique aspects of the market, and the way in which the public and private sectors should interact needs to be carefully engineered to fit the needs. The market has turned out to be much more complex than we ever imagined with a confusing array of choices and a wide range of unrelated actions and initiatives. At several points in the course of developing this book, at least one of the authors felt that this chaos and confusion was unmanageable and maybe even that it should not be managed, but allowed to grow under the influences of free market forces and the rules of business and commercial evolution. There is still a sneaking doubt in our minds that this might be correct, but having reached the end of the development process, we decided that there is sufficient insight, structure, and method that has been tried and tested that could help to support the market and bring traveler information systems to the masses.

Bear in mind that our intentions in writing this book are to share our thoughts, hopes, and concerns with you. We have no magic bullets and do not for one second pretend that we have all the answers. What we think we can do is capture and explain the essential facts of the traveler information business, provide some structured thoughts on how to make it better, and give you lots of food for thought as you define, develop, and apply your own particular approach. This is not a how-to guide to traveler information systems—you will still need to identify and consider the requirements of your organization and work out your own approach. You will still need expert advice and guidance from those who have been there before you and who have the knowledge of experience. In reading this book, however, you will gain an overall understanding of traveler information systems, be able to define and explain the key issues, and have a broad idea of several potential approaches.

1.3 Intended Readership

When we started to characterize the intended readership for the book, we realized that we had set ourselves a daunting task. The expression being all things to all people came to mind. We realized from experience that the basic premise of covering both the wide range of public- and private-sector interests and perspectives on traveler information systems would be a stretch. The range of interests and motivations is so varied and the backgrounds of the respective participants so diverse. Our continued optimism prevailed, and thinking through the task in a bit more detail we realized that the target audience for the book is indeed diverse, but there are a number of core principles covering a range of common ground that can form the basis for a cohesive backbone or skeleton, around which we can hang the specific materials required to cater to a diversity of interests. In fact, one of the primary values that we hope to bring to you is an understanding of the wide range of people and participating organizations that exist in an effective ATIS deployment. Because of this structure, we encourage all readers to read all chapters of the book. While certain chapters and sections will have more appeal to some readers, the full value of the knowledge and information contained within the covers will only be realized when taken as a single whole. The flow of the material begins with introductory sections that provide enabling information and knowledge for later chapters and concepts. The middle chapters are heavily focused on the introduction and explanation of certain key concepts and business mechanisms that we have identified in the course of our involvement in the traveler information business. Towards the end of the book, we explain how these concepts fit within the bigger public- and private-sector picture and explore the application of the principles before summarizing the key points of the materials.

The book is intended for a broad spectrum of readers who are interested in gaining a structured view of the world of traveler information systems—from college and university students widening their knowledge with respect to transportation, advanced technologies, and business approaches, to senior level practitioners seeking to blend additional insights and practical experience with that of their own, in the course of developing practical approaches to the planning, development, and deployment of ATIS. We also strongly recommend that you read all of the following sections, even the ones that describe other reader types. As we stated earlier, we hope to provide you with an appreciation and understanding of the range of people and organizations that could be involved in ATIS development and implementation. One of the essential features of ATIS is the need for cross-sector collaboration and partnership development across different types of organizations and communities. Reading through all of the following sections describing the intended readership for this book will help you to understand the range of potential participants and have at least an outline understanding of their characteristics and motivations. So please do not skip to the section that describes who you are, take a look at the other sections and get an idea of the other major players. There are also points woven into the fabric of each section that are relevant to your overall understanding of the field, which you will miss if you skip ahead.

Academics

Members of this audience group are assumed to be students or teachers in either transportation or business-related subject areas. Out intent for this group is to supplement and complement specialized reading and study related to the core subject matters, with broadening ancillary information that sweeps over the subject of ATIS. We hope that this audience group will be able to explain the overall context and nature of traveler information systems, understand the nature of the technologies, and grasp issues related to practical applications. Based on our introductory materials on concepts and terminologies, readers should be able to understand and evaluate the salient points related to the practical application of the technologies.

From a business student or teacher perspective, we hope that our discussion on the traveler information supply chain in Chapter 5, the use of business models in Chapter 6, and the proposed business-planning methodologies in Chapter 7 will provide a framework and context to stimulate further detailed work and research in this important area.

Researchers

For those readers engaged in various types of research—such as public policy, business, and social analysis research—we offer a starting point for further investigation. Our attempt to structure the world of traveler information systems into a value or supply chain may be a good framework for structuring research in several directions emanating from the traveler information systems common starting point. The relationship between traveler behavior and the availability of accurate and complete information on options and choices strikes us as a particularly fertile area for further research. The identification and selection of appropriate business models for engaging public and private resources might also offer some interesting opportunities for research work that builds on what we have presented here.

We hope that this group of readers leaves these pages with an overall sense of the major issues associated with traveler information systems planning, development, and deployment and equipped with some ideas on which to build additional work.

Public-Sector Transportation Professionals at National and Local Levels

Those readers that fall into this category are dear to our hearts. We are all of and from the transportation community and have deep roots in traffic, transportation, and intelligent transportation systems. While we have tried to ensure that this book is equally applicable and useful to the other audience groups, we have to confess that because of our backgrounds and experience, this material was originally intended for you. We realize, of course, that the successful application of information and communication technologies to traveler information requires the effective engagement of a varied range of participants drawn from other audience groups, hence the inclusion of these other groups in our description of the target audience. We firmly believe, however, that many of the actions and responsibilities associated with the subject area fall squarely on the shoulders of the transportation profession. It may not always be a lead role or responsibility, but due to the huge potential for influencing traveler choice and transportation-network usage, it is always an important one. We have divided this audience group into a selected number of smaller groups (transportation operations managers, transportation planners, traffic and transportation engineers) in order to refine our definition of objectives and aspirations.

Transportation Operations Managers

This group of readers will have roles and responsibilities relative to the operations and management of transportation facilities and systems. Freeway and tolled expressway, transit system, and incident-management operational-management professionals are some of the readers we expect to see in this category. For this group, we hope to illustrate how the private sector can be effectively engaged in the quest for better traveler information, supporting the win-win scenario where public transportation objectives are addressed by private-sector effort and investment while simultaneously creating and developing viable business opportunities. Our treatment of private-sector needs and motivations hopefully will shed some light on the factors to be taken into account when making an approach to the private sector. The definition of a basic supply chain, various business models, and an overall approach to planning a traveler information systems business should all be helpful in this respect.

This group should also be interested in the definition and understanding of the operational needs of traveler information systems, as this group will undoubtedly play a major role in satisfying them. Our explanation of the fundamental business processes at work within the overall context of the traveler information systems domain should provide the basis from which operational needs and plans to satisfy them can be developed. Our treatment of the development and operation of data-collection infrastructure and information processing facilities should be particularly useful and valuable to this reader group.

Transportation Planners

The promise of advanced traveler information services available on a region-wide comprehensive basis should be particularly exciting to transportation planners as they try to implement and manage their policies and goals. It seems to us that a great many of the transportation policy objectives that this reader group strives to achieve can be addressed through the widespread and effective application of traveler information services. Improving accessibility and mobility while minimizing the undesirable side effects of the transportation implementation and operations processes requires that travelers make smart choices with regard to their mode of travel, routes used, and timing of the trip. While many transportation planning and transportation economics professionals have discussed the concept of hard demand management measures to force such choices through the management of the cost of transportation, a more subtle and effective approach lies in the application and use of suitable information to influence, persuade, lobby, and cajole the traveler to make better travel decisions. The provision of high-quality, reliable traveler information at the right time, in the right location, in the right format, and at the right price, has a huge potential to impact the way in which people make travel decisions.

Unlocking this potential requires the adoption and application of a structured objectives-driven approach to the development and delivery of information services and products in order to ensure that the desired transportation policy objectives are being addressed and ultimately satisfied.

We hope that by reading this book, this reader group will be inspired to investigate the use of traveler information products and services as a primary tool in the efficient and effective attainment of transportation policy goals and objectives.

Traffic and Transportation Engineers

We hope that we have delivered to this audience group a good overview of the technologies that can be applied to the delivery of effective and useful traveler information to the traveling public within your jurisdiction or area of responsibility. While we have not attempted to provide a detailed how-to guide on the selection and application of information and communication technologies, we hope to raise your awareness to the point where you have a good understanding of the close relationship between traveler information and traffic management and are inspired to investigate ways in which transportation needs, issues, problems, and objectives can be addressed through the application of traveler information systems and technologies.

We want to highlight the importance of big-picture planning with respect to traveler information and traffic-management systems in order to maximize the effectiveness and efficiency of designs and implementations. The big picture in question should include other transportation agencies, organizations, and applications as well as the private-sector market participants that play in the wider domain of information and payment services.

An important aspect of the collaboration and coordination we hope to inspire relates to the joint development, implementation, and operation of sensor and communication networks for transportation-network monitoring and management. When the bigger picture is viewed instead of the narrow, bottom-up perspective of the single-application, single-project, single-organization approach, it is strikingly obvious that the data collection, communication, and, indeed, the information processing needs of traffic management and traveler information have a large area of overlap. Much of the data needed for effective management of the transportation network according to predefined performance measures can also be used to feed the traveler information services and products that describe the transportation networks operation and option to the users.

In some cases, the integration between traveler information and traffic management is so close that it is almost impossible to distinguish the two. Take the case of the delivery of travel time information over variable message signs at the roadside. Does the message on the sign constitute a traffic-management service or a traveler information service? The answer, of course, is that it can be both at the same time. The message on the sign could be viewed as a traffic-management strategy designed to encourage drivers to take an alternative route. It could also be viewed as a traveler information service designed to ensure that the driver has the best available information on which to base further travel decisions. While the phrasing or wording of the message can bias it towards either purpose, the fact remains that the message could be used for either.

The possibilities for unlocking this synergy and in particular, sharing the cost of development, deployment, and operations of information and communication technologies in transportation are fascinating, and we hope to convince this reader group to become aware of the opportunities, develop a strong interest in exploiting them, and be motivated to move forward and take the appropriate actions.

Nontechnical Decision Makers

This takes us back to the policy satisfaction possibilities of good, well-planned ATIS. We hope that by reading this book that this audience group will appreciate the possibilities and the potential that traveler information systems have and allow this knowledge to influence and affect their policy definition activities and attainment strategies.

The type of high-quality, ubiquitous traveler information service we define and describe in this book is capable of providing effective support and underwriting for a range of diverse transportation policy objectives—from road capacity maximization to modal shift.

We also hope to illustrate and explain the path from the definition of transportation policy objectives through the development and design of traveler information systems to the operations and management of the same. The definition and description of the traveler information value chain and the types of business models that can be employed will hopefully encourage policy makers to be proactive in the promotion and advocacy of ATIS.

We also hope to encourage the adoption of a needs-driven, policy-based approach to the development and design of such systems in order to ensure that public transportation policy objectives are effectively and efficiently addressed. From a more radical perspective, we hope to inspire the nontechnical decision makers who have control of funding and budgets as well as policy-setting responsibilities to insist that traveler information systems are developed and deployed in a structured, coherent manner that maximizes the chances of success while addressing policy objectives.

Information Service Providers

This is a book about two primary perspectives: that of the public sector and that of the private. We hope to foster a greater understanding and appreciation in both sectors by providing the information and knowledge in this book. We take a close look at the needs and motivation of both sectors in the hope that we can encourage a deeper understanding of the relative strengths, weaknesses, opportunities, and threats that are incorporated in both sectors. From the perspective of the private-sector information service provider, we hope to shed light on the drive and operation of the public-sector transportation community as it defines needs and problems and attempts to invest common funding for the greater good. Through the definition of the value chain and business model options for the traveler information process, we hope to highlight the possibilities for the private sector to collaborate with the public sector and coexist in a mutually beneficial manner. Examples and illustrations of potential frameworks for collaboration are intended to support this end. We also hope to provide some constructive and useful input to those private-sector players as they define their business approaches through market assessment and strategic business-planning activities.

Device Manufacturers

For device developers and manufacturers, we hope that this book will help them to identify and assess the full range of opportunities that ATISs present. We assume that this reader group will be engaged in the planning, development, and manufacture of consumer electronic devices for personal and in-vehicle applications. They may also be engaged in the development of devices that make use of one or more of the traveler information system enabling technologies we describe in Chapter 3. While it is highly likely that such organizations will have conducted a fairly sophisticated analysis of the size and nature of the markets they plan to address with their devices and products, we hope to provide input to such plans at various stages of their development. We hope that the detailed profile of the public-sector participants that we provide in the book proves to be of value to this reader group when refining plans for new product development and market assessment.

Investors and Potential Investors

We hope that this reader group will extend to cover private-sector investors who may be considering an investment in the development and delivery of traveler information services or products. Our definition and explanation of the traveler information supply chain and the various business models that might be employed as the basic structure of business relationships and operations should provide useful information to this reader group

Consultants

For this last private-sector reader group we hope to provide a general grounding in and understanding of the issues and opportunities associated with the planning, development, deployment, and operation of ATIS. There is such a large amount of new information associated with the subject as well as a high need for the application of new skills, knowledge, and experience that we think there is a strong market for consulting services. While the whole subject area is obviously a part of the traffic and transportation consulting services market, we believe that there is also a need for the application of additional professional skills, knowledge, and experience related to the information and communication technologies applied and the supply-chain management and business models utilized. We hope that the coverage we provide in these areas will enable professional services firms to identify the need for specialist consulting services and match these to their current and future capabilities.

1.4 Structure and Content of the Book

The structure of the book has been carefully developed to provide maximum support for our overall objective of introducing you to the world of ATIS and showing you some effective and practical ways to approach the subject. The structure is designed to provide a build effect, with early chapters paving the way by providing information and background required to fully appreciate later chapters. Starting with a basic definition of what we mean by ATIS, we move on to address the nature and characteristics of ATIS technologies, such as sensors, telecommunications, data storage, and information processing technologies. This is followed by a discussion of the public- and private-sector objectives that form the basis for sector interest in participation in the application area. This highlights the potential for intersector collaboration leading to the definition of an overall model that can be used to structure such collaboration. Since there are many possible configurations within which such public-private collaboration can take place, we then provide an exploration of the various business models that are possible, followed by coverage of some detailed models that have been applied in practice. One of the most significant features of ATIS is the huge potential for data and information processing sharing across ITS applications; therefore, we follow up on this with an exposition on how to ensure the appropriate degree of coordination between such applications. Finally, we summarize the essential points and provide some suggestions for further actions. The following is a description of what is provided in each chapter of the book. We have tried to supplement this with a brief commentary on what we expect you to take from reading each of the chapters.

Chapter 2: What Are Advanced Traveler Information Systems?

This is the foundation chapter of the book in which we have tried to set the scene and develop a common frame of reference for the subsequent chapters. Starting with a detailed definition of the phrase advanced traveler information systems, we provide an overview of what we mean when we talk about such systems, differentiating them from conventional information systems and pointing out the unique services provided by them to the traveler. To highlight the nature of traveler information systems at the next level, we then provide a description and definition of the various components that can be found in a typical ATIS, explaining the characteristics of the element and the function of the elements in the overall system. To conclude the chapter, we provide some brief descriptions of existing traveler information systems around the world, bringing the subject to life through the description and explanation of some real examples.

Chapter 3: ATIS Technologies

This part of the book provides another look at the components and elements that can be utilized to create an ATIS.

In order to provide a structure for the information on the various technologies, we introduce the concept of the traveler information supply chain, which stretches from the transportation network to the consumer. Utilizing this as a framework, we take each step in the chain and describe the various technology options that are available. This is a high level overview of the range of information and communication technologies that can be adopted and applied within the context of advanced traveler information systems. Our aim in this chapter is to make you aware of the range of possibilities that exist through the delivery of a structured overview. This will not give you the information you will need to develop and design ATIS but will help to define your needs and objectives alongside technical possibilities and capabilities. A major theme in our work in the intelligent transportation systems field has been the effective support of the iterative cycle between what you need and how it can be achieved. This chapter gives you some insight into how your requirements could be satisfied through the application of a range of information and communication technologies, products, and services.

Chapter 4: Public Objectives, Private Enterprise

Both the cultural differences and the commonalities between the public- and private-sector participants in advanced traveler information products and services are explored in this chapter. Starting with the public sector, we explore the types of roles and responsibilities that may exist in the various public-sector organizations and agencies that participate in ATIS planning, development, deployment, and operation. Building on this definition of who the participants might be, we then explore and discuss the typical objectives that the participants may have identified and defined as part of the justification for participation. This explores the motivation of the participants and the general goals they seek to accomplish through participation in ATIS.

The second half of the chapter goes through the same analysis and definition for the various private-sector participants, defining potential roles and responsibilities and likely objectives and motivations. The aim is to identify and contrast the cultural differences between the public- and private-sector participants, while also shedding some light on areas of potential synergy and collaboration, where there might be common ground.

We complete the chapter with a brief section explaining the differences between real objectives and metaobjectives, as we have labeled them. Metaobjectives are descriptions of potential solutions that are often used in place of the true goals and objectives for the project, system, or initiative. We include a short discussion on this as we have found that many public agencies define such metaobjectives, but fail to identify, describe, and confirm the real objectives for participation in ATIS. For example, an agency may decide that the goal is to have a traveler information system operational within, say, the next 12 months, rather than define the transportation policy objectives to be addressed by the system if it is to be considered successful by all users. We consider this point to be fundamental to successful ATIS planning, development, deployment, and operation. For success to be achieved and perceived, predetermined measurable goals must be established and agreed upon. While these could simply relate to the deployment of a system, a more powerful justification for involvement and investment lies in mapping the deployment back to transportation policy objectives.

Chapter 5: The ATIS Supply Chain

In support of our desire to provide a structured view of ATIS from both a technical and business point of view, in this chapter we define a high level process of the traveler information supply chain. The traveler information supply chain describes the overall process and the individual steps through which resources, such as money, time, hardware, software, data, and people are converted into the desired value to be delivered to the traveler. We define and describe six basic steps in the supply chain, explaining what activities take place within each step and how the six steps relate to each other in the overall process of the traveler information supply chain.

This information prepares the way for the discussion in Chapter 6, where we explore the potential roles and responsibilities of the public and private sectors in the supply chain. While the supply chain described is not sufficiently detailed to serve as a design template for an ATIS deployment, it serves the purpose of providing an overview of the major elements in the supply chain and acts as a vehicle for a structured discussion on the respective roles and responsibilities that might be assumed by each sector in the development and delivery of traveler information products and services.

We conclude this chapter with a short discussion on the application and use of supply-chain-management techniques. This makes the important point that the overall process or supply chain must be managed and operated in addition to the activities within each step or elements in the supply chain. Having recognized the existence of the chain and defined the relationships between each of the steps and the impact of activities in one step on the other ones, it makes sense to ensure that the overall process is being managed in addition to the individual steps. This is one of the valuable insights provided in this book. ATIS can and do function as a series of disconnected steps or activities, but they have the greatest chance of success and attainment of original objectives when managed as a complete process. Participants within each step may not have the motivation or resources to do this, so supply-chain management requires special consideration and treatment.

Chapter 6: ATIS Business Models

Utilizing the traveler information supply chain as was defined and described in Chapter 5, we explore the possible roles and responsibilities that participants may identify and adopt at each step in the supply chain. We begin Chapter 6 by providing our definition of a business model; this is to make sure that we are thinking the same thing when we see the term used in this chapter. Then we explore and discuss a range of basic business models that could be adopted in the planning, development, deployment, and operation of an ATIS. This results in a set of theoretical possibilities that explains the options available and how they might work. It was obvious to us, however, that additional detail to the basic models was needed in order to fully explain and describe some of the actual deployments in ATIS. While the basic models serve their purpose by illustrating the options available, we have adopted the use of our so-called enhanced business models to explain and illustrate some practical applications. The two sets of business models combined provide the reader with a comprehensive resource for exploring the possibilities and understanding the operation of practical deployments.

Chapter 7: Business Planning for ATIS

To complete the information set we started to develop in Chapter 6, we propose business-planning methodologies for the public and private sectors in this chapter. These explain the steps that are useful in the creation of a business plan for ATISs and how the business model fits within the overall planning context. The methodologies also indicate how dialogs between public and private sectors should be established and managed.

Chapter 8: How Traveler Information Systems Fit into the Future Big Picture

In this chapter we take a look at how ATIS fit within the larger transportation and intelligent transportation systems fields. Starting with a public-sector perspective, making use of the National Architecture for ITS, we illustrate the close connectivity and the data and information sharing that is feasible between ATIS deployments and other ITS applications. We pay particular attention to the relationships between ATIS and advanced traffic-management systems, both from a user service point of view and from an infrastructure perspective.

Moving on to the private-sector perspective, we examine the potential relationships between traveler information services and products and other information streams, such as sports scores, stock quotes, news, and weather.

We conclude the chapter with a brief view of potential future products and services that could emerge as ATIS evolve and become widely accessible.

Chapter 9: Summary and Conclusions

Like all good summary chapters, this one reviews the critical information delivered in the book. We also provide an overall summary of the advice we would offer to public- and private-sector participants in the realm of ATIS or one of the closely related fields.

1.5 A Vision for Future Advanced Traveler Information Systems

There is one more thing we have to address in this introductory chapter if we are to completely prepare the ground for subsequent chapters. We wrote this book in the hope of supporting some changes in the way participants approach and think about the development and implementation of ATIS. We believe these changes are necessary in order to enable us as a transportation and business community to make progress toward some ideal situation where traveler information is readily available to the public and fully supports positive travel behavior changes. We realize that much of the content of the book has been driven by our vision of this ultimate ideal and we believe that sharing it with you at this point in the book will be of great value in helping you to understand and appreciate later materials.

The vision we share for the future of traveler information has changed dramatically over the years in terms of how we expect it to be implemented and manifested in everyday life, but it has remained constant in terms of what services we expect to be delivered and the overall impact on the traveler. Maybe there is a lesson here in that due to the dynamic nature of information and communication technologies, defining the future in terms of how it will be done is a lot less stable than defining the future in terms of what you want in terms of products and services. In simple terms, the overall objective of an ATIS is to deliver and support a context within which there are no surprises for the traveler. Information accessibility should be such that the large majority of the traveling public has easy access to the full range and quality of information required to support informed and intelligent decisions at all stages in the trip and covering all modes. Let us explore the services we envision in support of this broad vision by taking a look ahead to some future time, say 2020, when at least some of the elements of the vision will be in place. Perhaps it will all happen well before then—who can tell?

We begin from the perspective of the end user (the traveler or consumer) for all the products and services that comprise the ATIS of the future. We describe and explain the products and services that will be available and provide an end results view of the future in terms of the impacts on the traveler. Then, as we wish to make some specific points regarding the operational aspects of the future system, we take a look through the eyes of the operators and managers of the future ATIS. This provides more of a back-office view of the future system by exploring the data-collection and information-processing facilities that are available and the transportation objectives that are being addressed.

1.5.1 Information User or Consumer Perspective

The year is 2020 and our future traveler information system makes use of multiple information delivery mechanisms to deliver traveler information to the traveler when required, in the appropriate format and with the necessary content and timing to effectively support travel decisions, such as choice of mode, choice of route, and timing of the journey. The system will also deliver sufficient information to enable hybrid or multimode trip options to be assessed in terms of time, distance, and cost parameters, alongside the more typical single-mode options. The information is available at each stage of the trip: prior to departure (short- and long-term advance trip planning information), at departure, during the course of the journey, and afterwards for analysis and cost-management purposes. The information is available in a wide variety of customizable formats including voice synthesized information bulletins, short messages, and sophisticated map-type displays. Each user can establish a profile stored in the system that describes information preferences, such as routes and modes of interest, timing for information delivery, preferred delivery channels, and formats. This enables a full account to be taken of the information user's needs and desires. Let us take a look at the use of traveler information at the various stages in the journey, namely the following:

- Pretrip;
- On-trip;
- Post-trip.

1.5.1.1 Pretrip Home- and Office-Based Information Access

Traveler information arguably is at its most valuable in terms of influencing behavior when delivered in the pretrip context. At this stage in the trip the traveler has the widest range of options regarding choice of mode of travel, choice of route, and the timing of the departure. It is also likely that the traveler or system user will have the best access to high-quality, high-capability input devices and information display facilities in the home or the office. Of course, in 2020 the technologies for mobile information access have advanced to the point where the quality and sophistication of in-journey information services is just as good, but the user still has more time and attention to spare on completing complex data input and information retrieval tasks in the pretrip context.

Given the importance of the pretrip stage of the journey, this is where the maximum range of information services is delivered. In our future vision for traveler information, the traveler accesses the information services within a flexible time period prior to the trip. Perhaps a few days before the trip, the traveler will refer to the information services for general information on route and schedule options. In the hours and minutes before departure the traveler may be interested in other information, such as the current status of transportation on the preferred route, the current cost of making the journey, and the levels of congestion that may be present. This would be the point at which nonrecurring or incident information is delivered enabling the traveler to reevaluate options in the light of current, accurate, and complete information on choices, options, and alternatives. Traveler information is also provided on a broad-spectrum basis covering all modes of travel including the following:

- Car (private car travel);
- Truck (short distance delivery vehicles and long-haul freight vehicles);
- Bus (fixed-route urban, intercity express and paratransit);
- Emergency vehicle (fire, police, and ambulance);
- Train (light and heavy rail, metro, suburban, and intercity);
- Airplane;
- Sea-faring ship;
- Bicycle;
- Foot.

The information also covers the complete door-to-door trip chain for people and goods from the origin of the journey to the destination. This is supported by a portfolio of multiple information services owned and operated by a range of operating entities. To ensure that the information is delivered on a consistent basis despite the variety of information sources, we anticipate the development and implementation of a number of key standards relating to data quality and information presentation with respect to traveler information. These would enable a consistent approach and the appropriate degree of commonality, while leaving each operator with the flexibility to provide additional value and differentiation. Operators would offer additional subscription and advertiser-supported services along with the requested or desired information. We assume that the traveler information service operators would supply these additional services on either an independent basis or in collaboration with one or more business partners.

1.5.1.2 On-Trip Information

In our future vision, traveler information is delivered in the course of the journey, through a range of mechanisms including in-vehicle, roadside, bus stop, rail station, airport, and other public places.

In-Vehicle

The in-vehicle component of the trip encompasses an integrated mobile information environment in the vehicle. Going well beyond the delivery of traveler information, this environment supports interaction between driver and vehicle as well as interaction between the driver and a range of service providers. The in-vehicle environment is original equipment, designed, developed, and implemented by the automotive manufacturer or by an automotive electronics supplier working for the manufacturer. The in-vehicle environment incorporates communication links with key systems in the vehicle and external systems via a wireless communications link utilizing either wide area wireless or short-range wireless communications technologies. Communications between subsystems inside the vehicle are handled by a wireless vehicle area network approach [1].

Depending on the manufacturer or automotive electronic supplier involved, a range of different architectural approaches to the in-vehicle system are being used. In some cases, the car radio/stereo/CD player approach is taken, where an imbedded, integrated unit is provided as factory fitted equipment. Use of the system and delivery of the information is constrained and customized for use in the vehicle.

In other cases, the handheld mobile telephone or personal digital assistant (PDA) model is adopted. This is where the device used for data input and information delivery and display can be decoupled from the vehicle and used outside of the vehicle environment.

Others make use of a third approach that is a hybrid of the first two. A component of the embedded system can be removed from the vehicle, extending the use of the system beyond the confines of the vehicle.

All three of these approaches are utilized in support of the delivery of a comprehensive list of traveler information services and additional services that straddle the boundaries with other application areas, including the following:

Vehicle-Based Payment Systems   The in-vehicle units used to support the traveler information services can also be used to support vehicle-based electronic payment systems, enabling drivers to make use of the equipped vehicle to pay for goods and services. These could be transportation fee payment transactions, such as paying for tolls or parking. They could also be more general-purpose transactions, such as paying for fuel, fast food, or other goods and services.

Vehicle-Based Information Services   Here again the presence of the in-vehicle equipment for the delivery of traveler information services enables other information services to be supported (or vice versa). This could be a range of services including yellow pages, car repair and servicing information, dynamic route guidance providing current driving directions to the destination, parking guidance, and reservation services enabling a suitable parking space to be identified, reserved, and located.

Non-Vehicle-Based Information Services In addition to the provision of information and payment services associated with the use of a vehicle, our future vision supports the delivery of information on a personal basis, with no requirement for a specially equipped vehicle. This addresses the other modes of travel and provides a choice for drivers who do not wish to buy vehicles equipped with the specialized information delivery and communications equipment required to receive the vehicle-based services. This also provides an alternative architectural and business model approach for the private-sector enterprises that wish to exploit the market opportunities. Here again there are technical options for the delivery of the information.

Personal Information Access The use of personal, portable consumer electronic devices, such as palm pilots, cellular telephones, pagers, and other PDA-type devices to access services and receive traveler information. Although this is mainly intended to address the needs of travelers in the course of the journey, it can just as easily overlap into the pretravel contexts of the home or office.

Remote Traveler Support This is also a method of providing access to traveler information and other services in the en route context, but it makes use of fixed infrastructure, such as information terminals and kiosks to provide access to the services and deliver the information required. The types of terminals and kiosks utilized may have a lot in common with the home or office terminals used in the pretrip context, but they are likely to be hardened or "ruggedized" for public usage.

1.5.1.3 Post-Trip Traveler Information

You are probably wondering about the possible value of post-trip traveler information. After all, the journey has been completed and there is no further need for current information on options and possibilities for travel. The end of one's trip, however, could also be viewed as the start of advance planning for the next one. Information gleaned and lessons learned from previous trips can be used to build up a knowledge source, or log, in support of even better future trip planning. In the case of business travelers, data in the log can also form the source data to drive management and reporting systems for traveling expense reimbursement and business financial-management purposes. Therefore, our future ATIS enables the traveler to store and retrieve traveler information describing past journeys in the post-trip context. Use is made of the same access arrangements and information delivery channels utilized at the pretrip stage of the journey, with interfaces available to support transfer of the data and information to external systems that handle additional information functions, such as the business financial management ones we mentioned earlier.

1.5.2   Operator and Manager Perspective

The future vision for ATISs we have defined thus far has taken a user service perspective. There is another complementary view of the future of ATISs that lies in the identification and definition of the infrastructure required to support the illustrated services. We review the various information and communication technologies that can be applied to ATISs in Chapter 3, but we have assumed in our vision that there will be comprehensive and complete coverage of the transportation network with respect to the sensor and telecommunications technologies required to provide the necessary data to feed traveler information systems. We also assume that additional data, such as public transit fare and schedule data is available in electronic format and is also used to feed the traveler information systems, which in turn have the data storage and information processing capabilities required to turn the data into meaningful information. Finally, we assume that a full range of information delivery channels in terms of the technical hardware, telecommunications, and information delivery devices is available and accessible to a wide cross-section of the public. These elements of the future ATIS are assumed to support a wide range of information services related to travel and the status of the transportation network.

If you are thinking that our vision for the future seems reasonable from a user service perspective but that the assumed infrastructure required is a long way beyond the current state of deployment, then you would be right on target. One of the primary motivations for writing this book is our common desire to stimulate progress towards the development of the infrastructure required to enable the services to be provided.

Another facet of our vision from the operations and management perspective is the availability of clearly defined and agreed upon public-sector requirements for ATIS activities and investments. We assume that the necessary requirements identifications, exploration, and agreement work has been carried out and that the development and deployment of the system is guided and influenced by a clear picture of the public-sector needs, issues, problems, and objectives to be addressed if the system is to be considered successful.

In summary, our operations and management view of the future world related to ATIS features requirements-driven investment in the data-collection and information-processing facilities required to effectively and efficiently achieve public objectives and support private enterprise.

Reference

[1]   Society of Automotive Engineers, "In-Vehicle Networks," Special Publications 1509, March 2000.

2

What Are Advanced Traveler Information Systems?

2.1 Learning Objectives

After you have read this chapter you should be able to do the following:

- Define the term *advanced traveler information system*;
- Explain the difference between data and metadata;
- List the different trip stages where traveler information can be delivered and utilized;
- Describe the TravInfo® ATIS application in the Bay Area, San Francisco;
- Describe the Vehicle Information Communication System (VICS) ATIS application in Japan;
- Describe the Trafficmaster™ ATIS application in the United Kingdom;
- Define the seven basic steps that comprise the traveler information supply chain or business process.

2.2 Introduction

The term *advanced traveler information system* is open to interpretation since it is a relatively unfamiliar term to most people. This chapter provides a definition and description and establishes the context for much of the remainder of the book.

In this chapter we will define and explain what we mean by ATIS by taking each word in the phrase and discussing it. We will explore the kinds of data and information that can be handled by such systems and show how resources are integrated and applied to the delivery of traveler information. Then we will put the entire expression back together again and offer our formal definition of the whole thing. We will conclude this chapter by describing some examples of existing advanced travel information systems from around the world.

Let us start by analyzing the expression *advanced traveler information system* word by word.

Advanced

What do we mean by the word *advanced*? There are many ways to collect data, process it, and disseminate traveler information, some of which require little by way of information and communication technology application.

For example, a telephone network or wireless radio facility connecting people in the field back to a central base can relay anecdotal data from the public and from a few spotter planes or helicopters where it is disseminated by TV or radio to the public. While this approach makes use of some information and communication technologies, we do not describe it as advanced for three reasons. First, the reliance on anecdotal, subjective data for supplying traveler information leads to a degree of unreliability and inaccuracy that damages the credibility of the information service. Advanced traffic and transportation sensors linked by appropriate communication technologies back to a central processing center can achieve much better results, albeit at a higher capital investment level. The second reason is that reliance on a high degree of manual intervention and human processing in collecting the data, turning it into information and disseminating it, keeps the operations and management cost of the data collection, information processing, and delivery process high. It does not take full advantage of advanced information and communication technologies. These include database management and delivery mechanisms that can fully automate the processing and delivery of information, reducing operating costs and providing greater product and service flexibility. Third, on their own such approaches to the collection of data and the dissemination of traveler information do not support the delivery of what we have defined as decision-quality information to the user of the transportation network. This goes beyond simply making the user aware of current travel conditions by providing a full set of customized and specific information regarding alternative modes, routes, and journey timings, enabling the traveler to improve the quality of travel decisions. This, in turn, supports the ability of the ATIS to influence traveler behavior toward the satisfaction of transportation policy objectives and goals. The availability of decision-quality data leads to the availability of decision-quality information, which in turn leads to higher quality travel decisions, helping to optimize the operational efficiency of our transportation networks and facilities.

Do not get us wrong on this one, we are not condemning or ignoring traditional, manual traveler information systems. In fact, most large systems in operation today have a significant manual component due to the lack of complete sensor or communication coverage, and we will illustrate this as we explain and describe ATIS implementations at the end of this chapter. What we are saying is that the manual approach by itself is not the ultimate goal and that, while it can be a stepping stone on the path to full automation, it is not adequate for future needs. Consequently, our focus in this book is on what we are calling ATISs, which take full advantage of information and communication technologies to automate the data collection, processing, and dissemination process.

Traveler

A fairly obvious definition of a traveler is one who travels. It seems trivial at first, but there is a bit more to it when you start to explore the term. It is also apparent that there can be a difference in perception with regard to the meaning of the word traveler as witnessed by a recent survey conducted by the Gallup Organization on behalf of the Intelligent Transportation Society of America. As part of the survey work, focus groups were held at a number of locations across the United States, to assess consumer opinions on the new three-digit dialing code (511) for traveler information in this country. One of the interesting suggestions from the focus groups was that the word traveler be avoided when developing marketing and promotional materials for the new number since it was perceived to mean a tourist, or someone who is not familiar with the geographical area in question. Tourists and visitors are obviously a major target for traveler information services as their unfamiliarity creates a strong demand for such services. However, there is a considerable level of investment required in order to establish and operate ATISs, making it vital that the target audience covers the widest spectrum of the traveling public in order to generate the desired levels of benefits and value. Other types of traveler—especially the local commuter who represents a significant proportion of the target market and generates a large percentage of the daily traffic in urban areas—must be addressed. It is apparent, therefore, that clear identification, definition, and understanding of the target audience are vital to the success of any advanced traveler information system. So, let us explore the meaning of the word *traveler*.

Travelers can have a range of characteristics according to the mode of travel chosen, the purpose of the trip being made, and the characteristics of the individual. This last point is sometimes overlooked, but it is vital to take full account of the capabilities of the recipient of the traveler information. This ensures that the information is delivered in the appropriate format, maximizing its effectiveness, and also ensures compliance with legislation, such as the Disability Discrimination Act in the United Kingdom and the Americans with Disabilities Act in the United States. Both require that account be taken of the physical challenges associated with travel and access to information services. These characteristics affect the information requirements of the traveler and the choice of delivery mechanism (e.g., vision-impaired travelers may express a preference for voice-based information delivery). They are an important element in the identification and definition of ATIS requirements.

Let us explore some of the different characteristics and contexts for travelers. Starting with the mode of travel, this could be one or more of the following:

- Private car driver or passenger;
- Transit driver or passenger;
- Truck driver;
- Taxi driver;
- Air pilot or passenger;
- Ship or ferry captain or passenger;
- Cyclist;
- Pedestrian.

There can be many possible purposes for a trip or journey, but we can combine them and broadly classify them as follows:

- *Work-related:* Work-related travel involves traveling either to and from your place of work or in the course of your work. This travel could be generated by your need to attend a meeting at a location other than your office or by virtue of the fact that your job incorporates the operation of a vehicle. By definition, all drivers and operators of transit vehicles, taxis, and trucks would fit in this latter category. Commuters, or those travelers shuttling between home and work, represent an important component of this trip-purpose group since much of our urban travel demand is generated by such travelers.

- *Daily life–related:* Daily life–related travel involves getting from one place to another in connection with daily life activities other than work. These might include shopping, making personal visits to friends and relatives, taking the children to and from school, and all the other activities that require mobility.

- *Leisure-related:* Leisure-related travel involves traveling because you want to. The journey itself may be the objective as much as the destination, or you may be traveling to a location where you will spend some leisure time. This category includes all tourism-related travel.

- *Emergency-related:* Emergency-related travel incorporates the need to travel to obtain emergency medical services, the need to evacuate due to natural or man-made disasters, and the need for emergency services to reach you.

Depending on the mode and purpose of travel, the needs of the traveler, especially the need for information, vary substantially. For example, it is likely that a commuter is interested in the travel conditions on a very specific set of routes and or modes with the objective of identifying the travel option that will produce the shortest or the most reliable travel time. The leisure traveler may be much more interested in context information regarding selected routes and information on opportunities to detour and perhaps spend money.

We also need to consider the full range of characteristics that a traveler may exhibit. As we shall discuss in later chapters, a needs-driven approach to the development, design, and deployment of ATIS is crucial to success, and so understanding the customer is very important.

Information

What do we mean by information? We carried out a quick Web search and came up with the following two definitions for information [1]:

1. Facts, data, or instructions in any medium or form;

2. The meaning that a human assigns to data by means of the known conventions used in their representation.

Information also reduces uncertainty by providing additional knowledge and adding to what the user of the information already knows.

There is a wide body of knowledge and research into the nature of information and how humans make use of and react to information. This ranges from information science to cognitive psychology studies. For the purposes of our book we will adopt the following definition, based on the formal definitions uncovered during the Web search: Information is data that has been subjected to processing to analyze the underlying patterns and trends to reveal facts that are useful to the user because they reduce uncertainty and shed light on a subject that is important to the user [1].

As we discuss in Chapter 6, the value of traveler information can be quantified rigorously, at least on the conceptual level, in terms of reduction of uncertainties related to rational decisions to be made by the traveler. This leads to another question: What is data? Referring again to the Department of Defense (DoD) dictionary Web site, they define data as the "Representation of facts, concepts, or instructions in a formalized manner suitable for communication, interpretation, or processing by humans or by automatic means. Any representations, such as characters or analog quantities to which meaning is or might be assigned [2]." Therefore, data is the raw material on which information is based; information goes beyond data by providing meaning to the user. We will reserve the term *information* in this book for advisories, messages, and processed data that can be easily understood by the intended user. We will use the term *data* to describe the raw inputs that are used to derive these messages.

Consequently, the term *traveler information* refers to information that has been derived from raw data concerning the transportation network or related entities by processing to analyze the underlying patterns and trends. This resulting information reveals facts that are useful and have value to the traveler because they reduce uncertainty in the transportation process.

This captures the objective, which is to provide the traveler with information products or services that have value and utility. The value attached to the product or service falls into one of two categories and depends on whether you are an end user (consumer) or a public agency trying to manage transportation. From the perspective of private providers, consideration of value must be given not just to a single end user, but to the entirety of all end users (including public agencies) that constitute the potential market for the product or service provided.

Direct Personal Value   This is the value the information has to the target customer. For example, travelers may value journey time savings, and having access to traveler information can lead to a time savings that could be translated into a value for the customer. Travelers may also value safety and security, so an information stream that includes weather conditions and safety-related information would also create a direct personal value for the traveler.

Indirect Community Value   This is the value to the whole community resulting from the expected behavior change that will be induced in the traveler as a result of having access to decision-quality traveler information. This also represents a step towards the achievement of one or more transportation policy objectives, such as smoothing out peak period travel demand or encouraging a shift from one mode of transportation to another. In this case the value lies with the expected benefit to the whole community, rather than the individual. This holds the most interest for the public agency or transportation manager.

We will return to the subject of valuing traveler information and the related data in Chapter 6.

Systems

When we researched systems in the same DoD dictionary used above, the closest we could find was a definition of system design [3]: "The preparation of an assembly of methods, procedures, or techniques united by regulated interaction to form an organized whole." Notice that the definition does not restrict systems to hardware, software, and data, but also includes methods and procedures. This fits well with our view of systems as being composed of hardware, software, data, and people.

We also looked to the Oxford English Dictionary and found the following definition of *system* [4]: "A group or set of related or associated methods, procedures, or techniques forming a unity or complex whole." Our preferred definition takes something from each of these definitions to yield the following working definition: "A group or set of related or associated material or immaterial things subject to regulated interaction forming a unity or complex whole." So, we have taken the phrase advanced traveler information systems apart to try to explain what it means in some detail. Of course, the full meaning is not entirely clear until we put it back together again and view it as a single expression. Drawing on the definitions for the words that make up the expression and synthesizing them, we arrive at the following definition of ATIS: "The systematic application of information and communications technologies to the collection of travel-related data and the processing and delivery of information of value to the traveler."

2.3 Systematic Application

As you will see later in the book (particularly in Chapters 6 and 7), we want to encourage you to take a systematic approach to the application of such technologies through the consideration of the subject as a system and through collaborative actions involving multiple public agencies and private enterprises. The term *ATIS* covers a wide range of information describing the transportation network and its current operational state. There are a number of dimensions to this information and we would like to explore these as a way of highlighting the practical meaning of the expression. It all revolves around four basic questions: what, where, when, and how?

2.3.1 What?

This aspect describes the information content itself. Scanning the numerous ATIS that are currently operational, we have put together a list of information content that is currently being provided to travelers:

- Travel times [e.g., road, bus, rail, light rapid transit (trams)];
- Transit schedule information;
- Transit fare information;
- Transit boarding and alighting point information;
- Locations of accidents and incidents;
- Description of accidents and incidents (including duration);
- Alternate routes and modes;
- Traffic speeds;
- Traffic delays;
- Queue lengths;
- Door-to-door multimodal journey times;
- Journey time reliability;
- Pretrip planning information;
- Travel-related weather conditions;
- Images (still images to full-motion, high-quality CCTV pictures).

The What? dimension also includes sets of parameters that describe the data. These are often referred to as metadata, as they represent data about the data rather than the data itself. Metadata can include the following parameters:

- Accuracy;
- Completeness;
- Reliability or confidence level;
- Timeliness (current status and future forecast);
- Format content and structure for; electronic messages;
- Accessibility;
- Relevance;
- Ease of use.

2.3.2   Where and When?

These dimensions address the location of the traveler information access point and the timing of the information delivery.

- Pretrip:
  - At home;
  - In the office;
  - Anywhere (via mobile devices).

- En route:
  - Roadside information on variable message signs;
  - In-vehicle information systems;
  - Private car;
  - Transit vehicle;
  - Smart bus stops;
  - Kiosks at transfer points and public locations.

- Post-trip (as described in Chapter 1):
  - At home;
  - In the office;
  - Anywhere (via mobile devices).

2.3.3 How?

This dimension deals with the way in which the data is delivered to the traveler. There are many delivery mechanisms available, including Internet access, wireless telephones, pagers, wireline telephones, and TV and radio. These delivery mechanisms will be described and explained in detail in the next chapter on traveler information technologies, but here is a brief overview of the range of major delivery mechanisms currently in use for traveler information purposes. You should recognize most of them from our vision for the future of ATIS described in Chapter 1.

2.3.3.1 Variable Message Signs (VMS)

Often referred to as changeable or dynamic message signs (CMS or DMS), these are signing units, or large display boards that employ information display technologies to impart information and instructions to drivers and other travelers. They can be located at key decision points in the travel process, such as important intersections along the highway, approaches to parking garages, or at multimodal interchange points, such as airport concourses, bus stops, or railway stations. While the information displayed on the signs is dynamic in that it can be changed quickly through a connection to a computer system, it is a shared medium that constrains the flexibility and customization of the message. All viewers of the sign see the same message at any given time.

2.3.3.2 Wireless Telephones, Pagers, and PDAs

The latest cellular telephones, pagers, and other handheld devices have advanced paging and even Internet access options available. These can be used to deliver traveler information on the move. Most of these devices have a keyboard or other input capability and a display that can support text messages. The more sophisticated devices have color screens and graphics capabilities, which support the delivery of graphical traveler information, such as maps and diagrams.

2.3.3.3 In-Vehicle Information Systems

A range of in-vehicle information units is emerging in global consumer markets in Japan, Europe, and the United States. These information delivery devices have display screens and input capability and may also incorporate speech recognition and speech synthesis. These have the potential to support the timely delivery of a wide range of information services, including traveler, weather, sports, news, and stock information. This type of service bundling is an important aspect of information systems, as it enables the content and format of information services to be tailored to the individual traveler's needs and lifestyle, maximizing the value to the traveler and optimizing the commercial viability of the individual services.

2.3.3.4   Dial-in Telephone or Interactive Voice Response

There are a number of interactive voice response systems available to support the delivery of traveler information through dial-in services. These are similar to the type of voice or touch key driven menu systems used by the customer service departments of many organizations. A user dials in to a predefined telephone number using a normal push-button, tone-dialing telephone as found in most homes and offices these days. An initial entry menu is spoken to the user by a prerecorded voice, and then through the use of the dial pad or by speaking commands, the user can navigate through various menus and submenus until the desired information is accessed. A good analogy would be the use of a Web browser to find your way around a specific Web site on the Internet. By clicking on links with your mouse, you can move from one page of the Web to the next. In the case of an interactive voice response system, you do not have a mouse or a display, so you have to press the buttons on the telephone or give speech commands to navigate.

2.4   Some Examples of Existing ATIS Implementations

In order to illustrate the topics addressed above, we provide descriptions of three ATIS applications or implementations from around the world. We have selected these on the basis of global coverage, with examples from Europe, the United States, and Japan. These examples also feature the collection of data from a number of sources internal and external to the operating organization, hence requiring data fusion to take place. Finally, we have selected examples that feature the use of multiple information delivery channels, such as dynamic message signs, Internet, and interactive voice response. It should be noted that the selection criteria used has resulted in the identification of three examples that predominantly feature traveler information related to roads and driving conditions. This is because most of the current deployments of ATIS that take data from multiple sources and feature collaboration across organizational boundaries have this focus. While we want to use these to highlight the major features of ATISs, bear in mind that the very best service to the traveler comes from systems that provide comprehensive information across all modes of travel and deliver information at all stages of the journey.

2.4.1 TravInfo®—San Francisco Bay Area

2.4.1.1 Overview

TravInfo® was one of the first large-scale ATIS projects in the United States. Centered around the San Francisco Bay Area transportation network, Trav-Info® began its life on June 1, 1993, as an ITS field and operational test or demonstration project sponsored by the U.S. Department of Transportation under the auspices of the Intermodal Surface Transportation Efficiency Act of 1992 (ISTEA). TravInfo® commenced the delivery of traveler information to Bay Area travelers in August 1996, operating a telephone information service as a public private partnership, collecting travel related data from the Bay Area transportation network, and providing traveler information to local travelers and consumers.

Public agency partners include the project managers Metropolitan Transportation Commission (MTC), California Department of Transportation (Caltrans) District 4, and the Golden Gate Division of the California Highway Patrol (CHP).

Under the terms of the partnership, the public-sector agencies provide a platform of traffic and public transit data that is available on an open-access basis to private-sector organizations. These private-sector partners, or information service providers, take the data and convert it into information products to be delivered to Bay Area consumers.

2.4.1.2 Products and Services Employed

TravInfo® utilizes a number of products and services as part of the overall data collection, processing and delivery system. Data is currently collected by a network of traffic sensors and from a variety of nonautomated systems.

There are traffic sensors operated by Caltrans, which are installed at a number of key locations on the Bay Area freeway network. These collect traffic data, such as vehicle speeds and traffic volumes and send the data back to TravInfo® Information Center. A variety of nonautomated data sources are also employed, including incident response reports from CHP's computer-aided dispatch (CAD) system and Caltrans' road closure operations center. An operator transcribes details of incidents from the CAD system to the TravInfo® system.

Data from the Freeway Service Patrol (FSP) is provided by a fleet of 50 tow trucks equipped with vehicle positioning systems. These form the MTC

Service Authority for Freeways and Expressways (SAFE) program's freeway service patrol and provide real-time road and traffic data on a vehicle probe basis. Specially trained cellular reporters who act as traffic probes by reporting current traffic conditions on an anecdotal basis complement this.

Additional information is provided by numerous Bay Area transit agencies. This tends to be predefined transit schedule and fare information rather than real-time current network status information.

Delivery of the information is achieved by TravInfo® public agency partners through a dial-in interactive voice response telephone system operated by a private-sector contractor. This enables travelers to dial in to one single central telephone number, then work through a series of menus to get the information they need. This is known as the Traveler Advisory Telephone System (TATS). It provides information to travelers through a touch-tone telephone service accessible from all Bay Area area codes. Callers are able to receive up-to-the-minute, route-specific information and are also able to connect to all Bay Area transit and ride-share providers. This is a good example of the type of information service bundling or grouping that is very important to the delivery of value to the user and the economic feasibility of future services. A TravInfo® Web site is also under development, which will enable traveler information users to access information in visual, graphical, and map-based formats.

On the private side, multiple delivery and distribution technologies and channels are utilized, including private-sector use of Internet, television, and radio broadcasts.

As you can see, the TravInfo® system comprises of a number of elements including data-collection infrastructure, data-collection processes, data processing and fusion, information creation and adding value, marketing, and distribution. As will be seen, this is a typical set of elements.

2.4.1.3   Interesting Features

One of the most interesting aspects of the TravInfo® system is the public-public partnership that has emerged as a result of the catalyst effect of the Federal Highway Administration funding for the field and operational test. Close collaboration between transit, planning, police, and traffic-management interests has bridged the gap between what are typically thought to be polarized self-interest groups. Another feature, related to TravInfo's genesis as a federally funded field and operational test, is the transition that has and is still taking place from research and development to a fully fledged business operation achieving well-defined public transportation policy objectives. Close observation of this process reveals a steady evolution from a technology and project focus to a business process and organizational structuring focus, as the system has become a mature component of the local transportation scene. This evolution has also featured a steady refinement of public-sector objectives over the life of the project to a clear focus on government assurance of the delivery of basic level services through widely available media—with other services being welcomed but not the main focus. For more information on this and other U.S. ATIS deployments, see [5, 6].

2.4.2   Vehicle Information Communication System—Japan

2.4.2.1   Overview

VICS is the national ATIS for Japan. As illustrated in Figure 2.1, VICS makes use of a combination of communication and information technologies to collect data on current road-traffic conditions and deliver traveler information to drivers whose vehicles have been equipped with suitable in-vehicle units. The VICS center is the hub of the operation: Data is collected and sent to the center for processing into the required traveler information. The center was established in July 1995, with initial coverage in the Tokyo region followed by the Tomei and Meishin Expressways in April 1996. Like Trav-Info®, data is collected through a combination of automated and manual means as depicted by the diagram.

2.4.2.2   Products and Services Employed

VICS employs a combination of automated and manual data-collection techniques to collect a variety of road-traffic data, including traffic congestion,

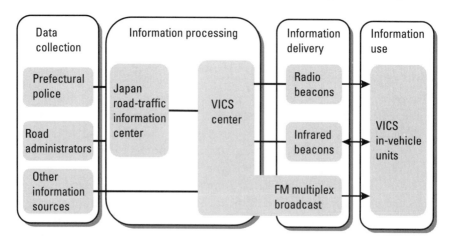

Figure 2.1 VICS flow diagram.

travel times, accidents, incidents, special event data, and road closures or traffic restrictions. Police and road administrators collect data from a number of sources including traffic-management systems. The data is collected at the VICS center and is delivered to drivers using three communications techniques and three types of in-vehicle information devices. Two of the communications techniques are spot or point wireless techniques (sometimes referred to as dedicated short range communications) where communication takes place with the vehicle at designated locations along the roadway, only when within range of a roadside transceiver or beacon. Radio-wave transceivers are utilized for expressways, while infrared beacons are utilized on trunk roads. The other technique is a wide area wireless technique known as FM multiplex broadcast. This utilizes existing FM stereo radio broadcasts to piggyback traveler information delivery to drivers across whole areas. The choice of technology is determined by the operating agencies for each type of road.

2.4.2.3   In-Vehicle Information Devices Employed

VICS delivers real-time information to drivers inside their vehicles, utilizing three different types of in-vehicle unit:

1. Map display;
2. Simple graphic display;
3. Text display.

Figure 2.2 illustrates the three different display formats used for each of the device types. On the map and graphic displays, the different colors signify levels of congestion. The text display presents a narrative describing traffic conditions and congestion. A large range of private-sector automotive electronics companies have developed and delivered in-vehicle information and navigation systems into the Japanese market. Consumers are offered a range of options in that they can choose between the three different display formats and purchase unit types accordingly.

2.4.2.4   Interesting Features

One of the most interesting features of the VICS system is the focus on the delivery of traveler information to drivers in vehicles within the context of a national business model. The data collected and the information processed by VICS has great potential to support the delivery of information over other channels, such as Web sites, telephones, or information kiosks. Until

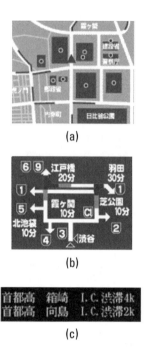

(a)

(b)

(c)

Figure 2.2 VICS in-vehicle information displays: (a) map display; (b) graphic display; and (c) text display.

recently, the public agencies and private companies involved in the development and deployment of the VICS system have chosen to ignore these potential delivery channels in favor of concentration on the in-vehicle units. This has enabled them to fully support a public-private partnership with a range of automotive electronic suppliers who have developed and delivered a wide range of in-vehicle units to the Japanese market. Sales reached more than 5.3 million units in March 2000, of which 1.8 million were equipped to make use of the VICS real-time information systems. For more information see [7].

2.4.3 Trafficmaster™—United Kingdom

2.4.3.1 Overview

Trafficmaster™ is an ATIS application covering the major motorways and many of the major trunk roads or arterials in the United Kingdom. In simple terms, the Trafficmaster system has three primary components: a network of traffic sensors, a communications network, and a range of in-vehicle information units. The traffic sensor network collects automated data on traffic conditions and then delivers information to specially developed in-vehicle information displays and delivery systems. The original Trafficmaster concept was promoted by a private-sector company known as General Logistics, PLC (later renamed Trafficmaster, PLC) under the provisions of the United Kingdom's 1989 legislation on driver information systems. This legislation, known as the Road Traffic (Driver Licensing and Information Systems) Act 1989, was originally intended to enable the Autoguide in-vehicle dynamic navigation program in London. Although the Autoguide program eventually foundered, the legislation enabled General Logistics to enter into a unique public-private partnership with the then U.K. Department of Transportation (UKDOT) for the collection of data and the provision of traveler information. A pilot license was granted by UKDOT in 1989 to cover the M25 motorway (London motorway ring), and this was extended to a full 15-year commercial license granted for England 1991 with separate licenses awarded for Wales and Scotland.

Trafficmaster commercial services were initiated in September 1990 and deployed on a phased basis across all the nation's motorways. At first, all motorways within a 35-mile radius of central London were covered. Then, in Phase 2 (spring 1992), all motorways in Southern England were covered. Phase 3 (spring 1993) extended coverage to motorways in northern England and Scotland. As of March 2000, the data-collection network was comprised of 2,400 infrared motorway sensors and 7,000 passive traffic-flow monitoring (PTFM) trunk road sensors, covering approximately 12,800 km of motorway and trunk road in the United Kingdom.

2.4.3.2 Products and Services Employed

The system utilizes two different types of traffic-sensor technology. For freeways, a network of traffic sensors are spaced at approximately 4 km intervals along the freeway network in the United Kingdom. The sensors are based on an infrared speed-measuring technology that enables the average speed of a group of six vehicles to be measured to an accuracy of plus or minus 8 km/hr. When the average speed drops below 48 km/hr, the sensors begin communication with Trafficmaster's National Traffic Data Center. The sensors are battery operated and wireless; wide area communication technology is used to communicate the data back to a central control center. This means that the sensors can be easily installed on existing bridges and gantries, requiring no hard wiring for power or communications. The wireless communications link back to control is achieved using a commercially available packet-data radio system, which keeps the cost of communication infrastructure to a minimum. From the National Traffic Data Center, the traveler information is delivered to the in-vehicle units using standard wireless paging techniques.

For arterials, it is not possible to erect sensors on overhead gantries. It is also not useful to have point data on the traffic conditions. In arterial traffic conditions, the point data varies so widely that it is not possible to obtain a representative assessment of traffic conditions from point sensors. Instead, link or area sensors that measure travel times along links in the highway network are required. Trafficmaster adopted a different technology for arterial roads, known as passive target flow measurement (PTFM). PTFM uses a combination of infrared cameras and image processing algorithms to automatically read the center digits from the license plates of passing vehicles. These are then matched at subsequent sensor locations downstream and average travel times are calculated for each road section of approximately 7 km. More than 7,000 sensors and transmitters have been installed every 3 to 7 km on both verges and bridges across the country. Specially developed by Trafficmaster, the network monitors journey times on arterials across the country and can also predict future journey times. The system allows Trafficmaster computers to predict when traffic is at its worst and therefore recommend the best time to travel. Computers at each site continuously transmit this information back to Trafficmaster's National Control Center, where it is processed and transmitted. The Trafficmaster sensor network is the most comprehensive, nationwide, live traffic-monitoring network in the world. In addition to providing information on traffic conditions, it also provides the motorist with real-time information on journey times on over 95% of the country's main roads. (Future enhancements will enable the system to predict peak traffic periods and allow the traveler to determine the best time to travel.)

On the information delivery side of the business, Trafficmaster has developed and delivered a range of in-vehicle, pretrip, and handheld units to make use of the real-time information feed from the extensive sensor network:

- *Trafficmaster freeway.* This is a stand-alone unit that utilizes speech synthesis to provide live traffic information and advance warning of delays up to 16 km ahead of the vehicle.

- *Traffic alert 1740.* This unit is designed to operate with Global System for Mobile (GSM) cellular telephones from a range of vendors. It uses the cellular telephone to bring the real-time traffic feed into the vehicle, providing live location-specific warnings on traffic hold-ups. In a similar way to the freeway unit, drivers are alerted to congestion up to 16 to 20 km ahead of their position. Drivers can then dial a dedicated telephone number to hear spoken information and additional details on the nature of the congestion. The user can also choose to talk to a live traffic information advisor on the same line.

- *Trafficmaster YQ.* This is the most sophisticated of the in-vehicle devices provided by Trafficmaster. It displays a map of the roads covered, indicating position, direction, and length of delays. The information is updated automatically every 45 seconds, 24 hours a day, and provides drivers with a built-in pager service at no extra cost. Figure 2.3 shows a YQ unit.

Trafficmaster markets and distributes all three in-vehicle units to U.K. consumers.

Additionally, the company has developed a range of in-vehicle units in collaboration with a number of automotive manufacturer partners. These

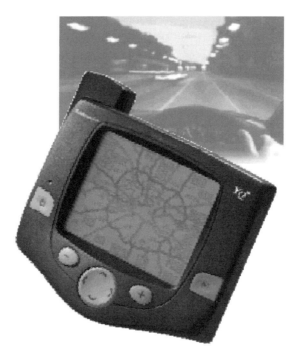

Figure 2.3 Trafficmaster YQ in-vehicle unit.

devices are customized to specific vehicles; partners include BMW, Jaguar, Citroen, and General Motors. More recently, Trafficmaster has developed and implemented a Web-based access channel or information delivery mechanism that enables subscribers to access traveler information from home, office, or mobile Internet terminals. Figure 2.4 shows one of the pages from the Trafficmaster Web site.

2.4.3.3   Interesting Features

Trafficmaster is one of the very few applications of advanced traveler information technologies where the data-collection network has been developed and established ahead of the delivery systems and devices. Looking back over the history of this enterprising company, it can be seen that initial investment in the data-collection infrastructure was very high in relation to the low numbers of people using the information services and buying or renting the devices. This initial weakness has turned into a unquestioned strength as the company now exploits the availability of a reliable, low-cost, automated data stream as the core for a wide range of information delivery services and in-vehicle devices.

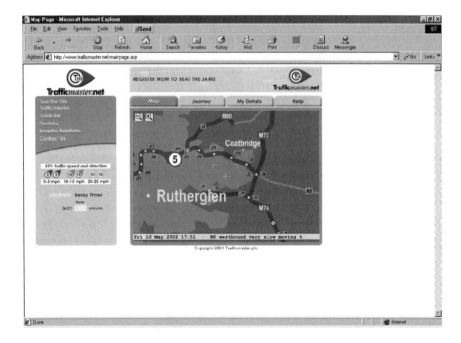

Figure 2.4 Trafficmaster Web site.

Trafficmaster is also unique as a privately owned and operated advanced traveler information system. Through a public-private partnership with the UKDOT [now known as the Department of the Environment Transport and the Regions (DETR)], they have financed and developed an entirely private data-collection infrastructure and operation. Trafficmaster is expanding its operations into multiple countries in Europe and the United States, though the approach taken in each country differs at least somewhat from the U.K. example. For more information, see [8].

References

[1]   Department of Defense Joint Publication 1-02, "DOD Dictionary of Military and Associated Terms," available at http://www.dtic.mil/doctrine/jel/doddict/data/i/02563.html.

[2]   Department of Defense Joint Publication 1-02, "DOD Dictionary of Military and Associated Terms," available at http://www.dtic.mil/doctrine/jel/doddict/data/d/01450.html.

[3]   Department of Defense Joint Publication 1-02, "DOD Dictionary of Military and Associated Terms," available at http://www.dtic.mil/doctrine/jel/doddict/data/s/05105.html.

[4]   *The New Shorter Oxford English Dictionary,* Volume 2, N–Z, Oxford: Clarendon Press, 1993, p. 3193.

[5]   "ATIS U.S. Business Models Review," prepared for U.S. Department of Transportation, ITS Joint Program Office, HOIT-1, Washington, DC 20590, Rick Schuman, Eli Sherer, November 15, 2001.

[6]   511 Case Studies, San Francisco Bay Area, Metropolitan Transportation Commission, U.S. Department of Transportation, April 5, 2001, available at http://www.its.dot.gov/511/Travinfo.htm.

[7]   "The Future Is Now, ITS Technologies to Ease Congestion and Improve Safety," *Japan Auto Trends* (Newsletter of the Japan Automobile Manufacturers Association), Vol. 3, No. 3, September 1999, available at http://www.japanauto.com/autoTrends/detail.cfm?id=67.

[8]   Trafficmaster, http://www.trafficmaster.co.uk.

3

ATIS Technologies

3.1 Learning Objectives

After you have read this chapter you should be able to do the following:

- List information and communication technologies applicable to ATIS applications;
- Define an operational concept for harnessing a group of such technologies.

3.2 Introduction

Cellular telephones, Internet, in-vehicle information systems, smart bus stops, and roadside dynamic message signs are just some of the technologies that can be applied to ATIS. In order to fully understand the nature of ATIS, it is necessary to have an overview and appreciation of the range of technologies available and of the appropriate applications. There are four major groups of information and telecommunications technologies that can be applied to the establishment and operation of ATIS:

- Sensors and surveillance;
- Telecommunications;
- Data and information processing;
- Information display and delivery.

This chapter provides a coherent introduction and overview of these four groups of technologies, which will enable a better understanding of their capabilities and usefulness with respect to traveler information. On their own, these technologies do not usually provide the benefits and improvements we seek in transportation and information systems. Simply having a wireless communication system on its own would not enable the delivery of effective traveler information, since there would be no data-collection or information processing capabilities. However, when these are configured into groups of related technologies, as specific products aimed at providing specific services and functions, the benefits become tangible and measurable. Thus, the world of ATIS involves the application of information and communication technologies in combinations that address needs and provide desired services and features. It is not our intent to provide a complete catalog containing every technology that could possibly be applied to traveler information systems. Nor is it our intent to conduct a comprehensive exposition on exactly how the technologies we do mention actually work. Our aim is to provide you with a reasonable understanding of what the technology possibilities are and give you a basis for further exploration into the wider world of information and communication technologies. We have also developed a specific focus on what services can be supported by the technologies and exactly what benefits they can provide for the traveler and the traveler information system operator. Consequently, our coverage of the technologies available will be at a fairly high level. We face a challenge in this respect, since the subjects of data sensors, information systems, and telecommunications represent a wide area of development and application beyond transportation. In order to keep our focus on traveler information as a specific application, we will provide an introduction to each topic as well as additional references and bibliographic information that you can make use of if you want to pursue the matter and learn more about particular technologies. To begin our exploration, Table 3.1 provides a complete summary of all the technologies we are going to describe. We have grouped them into the four major groups and subdivided them into taxonomies of technologies. We will follow this as an outline for our technology description sections. Table 3.2 maps the technologies to the type of data that can be collected by each.

3.3 Sensors and Surveillance Technologies

These technologies are incorporated and combined to form the data-collection infrastructure for ATIS. They are at what could be considered

Table 3.1
Summary of ATIS Technologies

Sensors and surveillance	Inductive loops
	Piezo sensors
	Radar
	Laser
	CCTV
	Automatic vehicle location
	License-plate readers
	Smart cards and other ITS
	Passenger counters
	Probe data-collection technologies
Telecommunications	Cellular wireless
	Wireless application protocol
	Broadcast radio and TV
	Bluetooth
	Copper wireline
	Fiber optics
	Dedicated short-range communications
Data and information processing	Data warehousing
	Data mining
	On-line analytical processing
	Voice processing
	Speech recognition
	Internet
Information display and delivery	Emergency call boxes
	Kiosks and smart bus stops
	Dynamic message signs
	In-vehicle information systems
	Personal information devices
	In-home or office-based delivery systems

Table 3.2
Data Collection Capabilities of Various Sensor and Surveillance Technologies

Technology	Spot speed	Volume	Class	Travel time	Queue lengths	Incident type	Origin/destination	Transit schedule-adherence data	Transit boarding- and alighting-point data	Locations of accidents and incidents	Description of accidents and incidents	Door-to-door multimodal journey times
Inductive loops	●	●			●					●		
VIDS	●	●			●					●		
Piezo sensors		●	●									
Radar	●									●		
Laser		●	●			●						
CCTV						●				●	●	
Automatic vehicle location				●			●	●	●	●		
License-plate readers				●			●			●		
Smart cards and other ITS				●			●					●
Passenger counters		●					●		●			
Probe data-collection technologies				●			●	●		●		

as the front end of ATIS as they are typically deployed around and on the transportation network, providing the raw data that goes into the information processing facilities. Sensor technologies are used to collect the raw data we need to describe and characterize the current conditions on the transportation network. Many of the devices, products, and services are capable of providing data for multiple applications, such as traffic management, incident detection, transit management, emergency management, as well as traveler information. This is one of the reasons why it makes the most sense to develop and deploy a traveler information system within an overall future big picture or system architecture. We will discuss this in more detail in Chapter 7, when we look at the larger context within which traveler information systems are deployed and operated.

Taking a look at the range of sensor technologies currently available and in use for transportation data collection, we find that they fall into one of two main groups: point sensors or area sensors. Point sensors, as the name suggests, focus their attention on either a single point on the transportation network or a relatively small, well-defined area or footprint of the transportation network. They enable the collection of instantaneous data regarding the vehicles that pass a particular part of the transportation network and lend themselves to the determination of temporal averages, such as time mean speed of passing vehicles.

Area sensors are capable of measuring and monitoring a wider area of the transportation network, such as a complete section of highway or a defined subsection of the transportation network, such as a zone. These sensors allow us to capture a snapshot of travel conditions on the network across the monitored zone. They lend themselves to the determination of geographical averages, such as average travel time within the zone or space mean speeds of vehicles. Figure 3.1 illustrates the difference between point and area sensors within a transportation-network context.

Travel times form the basis of the information that the user really wants in that relative travel times form the basis for route, journey timing, and mode choices. Predicted travel times based on trend and historical data can also be of considerable value in making travel decisions. Unfortunately, due

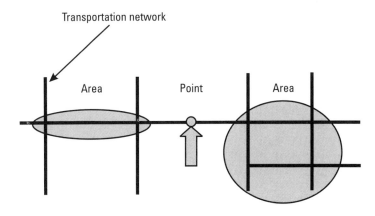

Figure 3.1 Point and area sensors.

to the method by which journey times are determined, they will always be slightly behind real-time conditions. This is due to the inherent latency in the calculation. In order to determine journey time, it is necessary for the vehicle to have completed the journey. Depending on the spacing of the journey time sensors, this delay could be anywhere from 1 to 30 minutes. Speed data, on the other hand, is almost instantly available and records the instantaneous speed of a vehicle passing a point. This is useful as a rough guide to travel conditions on freely flowing highways, but not for congested arterials and urban surface streets due to the wide variation in instantaneous vehicle speeds past the point in the highway where the measurements are taken.

3.3.1   Point Sensors

There are a number of technologies that could be of interest for traveler information systems application. In line with our overall philosophy for this book, we have not attempted to provide a comprehensive catalog of point sensors. Instead, we have selected a few examples that will illustrate the range and provide an idea of what can be achieved with sensors of this type.

3.3.1.1   Inductive Loops

Inductive loops have been in use of traffic counting and vehicle sensing for traffic control for more than 50 years. They work on the basic principle of inductance associated with an electromagnetic field and deliver accurate and reliable data for traffic and transportation purposes. Slots are cut into the road surface in a rectangular, diamond, or chevron configuration. Copper wire is inserted into the slot and wound around the configuration a number of times. An electric current is applied to the wire loop, inducing an electromagnetic field just above the road surface. Vehicles passing over the loop site cut through this field and generate a voltage difference in the wire, which is detected by monitoring and amplification circuitry and processed into traffic data. Note that single loop configurations can derive volume, occupancy, and classification, while a double loop configuration can directly measure the spot, or instantaneous speed of vehicles.

3.3.1.2   Video Image Detectors

Video image detectors (VIDs) perform the same function as inductive loops, without the need to embed sensors in the highway. Closed-circuit television (CCTV) cameras are installed on poles adjacent to the highway and are focused on the target area of the road. Sophisticated image processing algorithms process the video signal containing the road scene, detecting and counting vehicles passing through the target zone. The exact technique for processing the signal varies from vendor to vendor, with some models analyzing every pixel in the video image to varying degrees of resolution and others analyzing a thin line of pixels across the center of the road scene. The former approach enables vehicles to be tracked through the road scene also.

3.3.1.3 Piezo Sensors

These sensors take advantage of an interesting property of piezo crystal materials; that is, when physical pressure is applied they generate an electrical voltage that is proportional to the pressure. This gives them the added advantage of being able to detect the weight of a vehicle by converting mechanical stress or strain into proportionate electrical energy. Piezo sensors are thin strips of piezo electric material that are either laid on top of the road surface or in a slot like the inductive loops described above. Like the loops, monitoring and processing software is used to convert the voltage levels into traffic data.

3.3.1.4 Radar

There are two primary types of radar (radio distance and ranging) technologies employed in data-collection applications: Doppler radar and true-presence radar detection technology. Doppler radar is used to measure the speed of moving vehicles by analyzing the way in which a beam of microwave energy is altered as it reflects from a moving vehicle. This is possible in a portable format in the form of the handheld speed detector guns favored by enforcement agencies worldwide, or as permanently mounted roadside equipment linked by telecommunications facilities back to a data center or transportation-management center. The second variety, true-presence radar detection, has been implemented in the form of a traffic data-collection system known as Road Traffic Monitoring System (RTMS) and can detect traffic and collect the data parameters from both static and moving traffic [1]. It makes use of microwave radar to collect the same type of traffic data as inductive loops. A miniature radar is utilized to transmit a low-power microwave signal of continuously varying frequency in a fixed fan-shaped beam. The beam paints a long elliptical footprint on the road surface, monitoring as many as eight lanes of traffic simultaneously in the side fire configuration shown in Figure 3.2. Any nonbackground targets, such as vehicles reflect the signal back to the RTMS unit where the targets are detected and their range measured. Figure 3.3 shows an RTMS unit installed in a roadside setting.

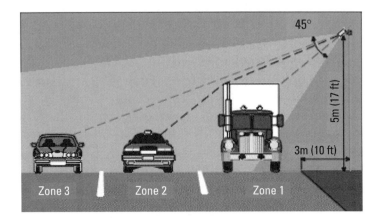

Figure 3.2 RTMS coverage. Used with permission of EIS Electronic Integrated Systems, Inc., July 2002.

Figure 3.3 RTMS unit installed. Used with permission of EIS Electronic Integrated Systems, Inc., July 2002.

3.3.1.5 Laser

Laser (which is an acronym for light amplification through stimulation and emission of radiation) technologies are utilized in a similar manner as radar except that a concentrated, coherent beam of light is used instead of microwave energy. One of the best examples of the use of laser technology for transportation sensing purposes is the autosense sensor [2], which makes use of a scanning laser beam to count and classify passing vehicles. It scans the roadway by taking 30 range measurements across the width of the road at two locations beneath the sensor. Each set of 30 range measurements forms a line across the road with a 10 separation between lines. At a mounting height of 23 ft, a 10 separation equals 4 ft between lines, as shown in Figure 3.4.

When a vehicle enters the beam, the measured distance decreases and the corresponding vehicle height is calculated using simple geometry. As the vehicle progresses, the second beam is also broken in the same manner. The unit calculates the time it takes a vehicle to break both beams, enabling the speed of the vehicle to be determined. Consecutive range samples are analyzed to generate a profile of the vehicle in view, as shown in Figure 3.5. This

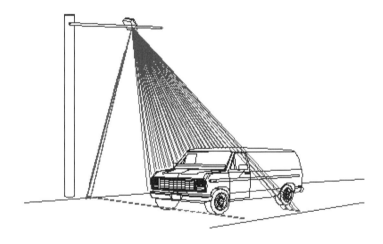

Figure 3.4 Autosense coverage. Used with permission of Schwartz Electro-optical, Inc., July 2002.

Figure 3.5 Autosense output profile. Used with permission of Schwartz Electro-optical, Inc., July 2002.

vehicle profile is then processed by the sensor to classify the vehicle into 13 different categories.

3.3.1.6 Closed-Circuit Television

CCTV for traffic and transit surveillance is one of the most popular applications of sensor technologies. It is so called because the TV signal is not broadcast over the airwave for any receiver to pick up and view (like commercial TV), but piped straight into a high-capacity communication network for distribution to a closed community of viewers. The use of near-broadcast-quality TV cameras and lenses, coupled with high-speed, high-capacity wireline communications enables transportation networks to be monitored continuously by both manual, human operator monitoring techniques and through the application of automated image processing techniques. The light that forms the image of the road scene is focused through the optical lens on the front of the camera on to a charge coupled device (CCD) inside the camera. This captures a snapshot of the road scene and translates it into a digital signal that is processed and then relayed over a telecommunications facility to a central traffic, transportation, or traveler information center. Depending on the video standard employed, this process is repeated 25 or 30 times per second, delivering a stream of snapshots that we see as a moving image when they are displayed on a TV or video projector. In the case of automated image processing techniques, sophisticated software is utilized to process the incoming video signal and detect or recognize certain predefined alarm or alert conditions, such as stationary vehicles or traffic flows above a threshold. It should be noted that we have defined CCTV as a point sensor since the cameras are located at specific points along the highway, but it could also be argued that they are area sensors since each camera covers a section of highway.

3.3.2 Area Sensors

3.3.2.1 Automatic Vehicle Location

There is a family of automatic vehicle location technologies that can play a valuable role in the collection of data from transportation networks. They include the following:

- Radio triangulation;
- GPS;
- Wireless transceiver tags;
- Cellular telephone.

Radio Triangulation   When a vehicle is fitted with a radio transmitter, a suitably placed radio receiver can receive the signal from that transmitter. Since we know how long it takes for radio signal to travel through the air (the speed of light, which is about 300,000 km/s), if we send a message that contains the time the signal left the transmitter in the vehicle and note the time when it is received at the receiver, we can convert the time difference into a distance. If, as shown in Figure 3.6, we have another two suitably located receivers, we can determine the distance from these receivers to the vehicle transmitter and through simple geometry locate the vehicle. Of course, you have to be able to access a whole network of receivers to be able to locate vehicles over a wide area. The U.S. Coastguard used to own just such a network. The system was called LORAN–C [3] and used low-frequency radio to locate coastguard vessels for a number of years until a better and more efficient technology known as GPS came along.

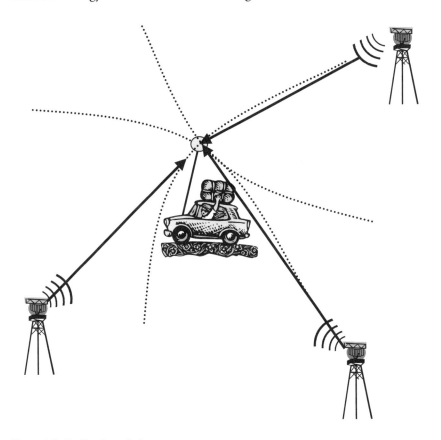

Figure 3.6 Radio triangulation.

Global Positioning System (GPS) This utilizes the same basic triangulation principle described above but does it in three dimensions using a constellation of 24 Earth-orbiting satellites that broadcast radio signals towards the Earth. The position of each satellite with respect to the Earth is known with great precision, thanks to a network of satellite ground stations and our understanding of the dynamics of orbits. A special GPS receiving unit can receive signals from a group of three of these satellites and determine its location anywhere on the planet. For ITS applications, such as automatic vehicle location, accuracies of approximately 10m are possible if three satellites are in line-of-sight simultaneously.

Wireless Transceiver (Toll) Transponders These are small boxes no bigger than a bar of soap that can support two-way communication with suitable roadside transceiver equipment [4]. They are often referred to as toll tags and form one of the key components of the electronic toll collection systems that have become prevalent in the United States, Europe, and Southeast Asia over the past 15 years. A unique vehicle identification number can be embedded in the tag and transmitted to the roadside infrastructure whenever the vehicle passes the equipped point. Of course, the same techniques can be applied to the identification of transit vehicles and rolling stock at various points on the transit or rail network, providing the basic data for vehicle tracking and fleet monitoring. The current version of this technology in use today has been specially developed and adapted for transportation use. It is likely that future versions of these technologies will converge with wireless local area network technologies, such as the IEEE 802.11b standard, to make use of widely available commercial technologies for short-range vehicle-to-roadside communications.

3.3.2.2 License-Plate Readers

These use a combination of video image processing and character recognition technologies to detect and read license or number plates attached to vehicles on the highway and use number matching software to determine travel times [5]. A typical system consists of video cameras, infrared light source, and a roadside processor. Typically, an infrared light source is utilized to illuminate the target area of the road. When a vehicle passes through the target area, the reflective sheeting of the license plate is illuminated by the light source (day or night), enabling the system to locate the license plate in the video image of the street scene. Character recognition software is then used by the roadside processor to identify the characters on the plate for subsequent entry into a central computer database over a communications link.

Further downstream a second installation reads the license plate again and the numbers are matched by the central computer database. The time taken for the vehicle to travel from the first installation to the second one is also logged as the travel time for that vehicle. The results from a large number of vehicles sampled in this manner are treated statistically to derive average travel times. Figure 3.7 shows a license-plate reader installed in a roadside setting. As we discussed at the end of Chapter 2, Trafficmaster™ in the United Kingdom makes use of this technology as part of their data-collection infrastructure and refers to it as PTFM.

3.3.2.3 Smart Cards and Other ITS as Data Collectors

Strange as it may seem, one of the most effective area-wide sensor techniques for transit systems is actually an electronic payment system. The kind of common stored value payment systems utilized in today's more advanced transit systems can be a terrific source of sensor data. Since the electronic payment system is capable of collecting details on the user's use of the system and preferences, this data could be used as a source for traveler information data. This leads to an interesting notion. We cannot characterize an electronic payment system as simply a sensor technology or data-collection infrastructure for traveler information. They are, in fact, systems or applications

Figure 3.7 License-plate reader. Used with permission of Pearpoint, Inc., July 2002.

in their own right with a defined set of objectives and features that are not directly related to the needs of traveler information systems. Yet, they have the capability to be a rich source of data for traveler information systems. This kind of synergy is exactly why we need to put the development and deployment of information and communication technology applications to transportation, into the context of a future big picture or system architecture. Further analysis of the relationship between the two systems will reveal that they are in fact subsystems within a larger ITS application with common data needs and strong operational ties. As an additional example, the fare structure and payment tariffs from the electronic payment system would be a key set of data for input into the traveler information system.

3.3.2.4   Passenger Counting Systems for Transit Vehicles

This is another example of a stand-alone system with distinct objectives and features being useful as a data source for traveler information systems. Advanced passenger counting systems incorporating sensors to detect passengers entering and leaving transit vehicles, such as buses and trains can supply loading or utilization data that can be used as input to traveler information systems. Such systems make use of laser, infrared, or light emitting diode (LED) technologies to detect passengers as they cut across a beam of light upon entry to and exit from the transit vehicle. There is, of course, the possibility that an electronic payment system incorporating a smart card used to register passengers as they enter and leave the vehicle could provide similar data, as discussed previously.

3.3.2.5   Probe Vehicle Data-Collection Techniques

A very powerful area-wide approach to the collection of data about the current operating conditions on transportation networks can be supported through the use of probe vehicle data-collection techniques. These include the use of a range of vehicle location technologies to track a sample of vehicles around the transportation network, enabling average travel times and delays to be determined. While different location and communication technology combinations are utilized, the same general principle is applied. Vehicle locations are sampled for a selection of vehicles on the transportation network. These are used as input to a software system that matches readings for the same vehicle at different points in the network, verifies the validity of the data, relates it to a representation of the transportation network, then aggregates the data into average travel times for each link in the network.

One such technique employs the wireless transceiver tags or license-plate readers described previously. As the tags are recognized at various points on the transportation network, the vehicle identities are recorded and processed to establish the progress that sample or probe vehicles are making through the network. Another version of this technique involves the use of cellular telephones as probes. There are a number of location techniques that can be used to determine the approximate location of a cellular telephone, including simple triangulation and more advanced signal mapping and processing approaches. Some techniques enable the location of the cellular telephone to be determined whether the user is engaged in a call or not, while others require the telephone be in use before location can be established. Some emerging techniques enable the location of the cellular phones to be established through the use of equipment and software at the cell tower or base station for that cell, avoiding the need to transmit the cell phone location over the air waves; others determine the location in the vehicle, requiring this to be combined with a wireless communications technology for transmittal of the location data back to a central location. At the time this book was being developed, the cellular wireless industry was considering the addition of a GPS chip to cellular telephone handsets in order to provide handset location data as the basis for a range of location-specific services including payment and emergency assistance applications. This would place the location determination technology in the handset, rather than in the network as featured in the other techniques.

The use of probes for data collection offers the highest potential of any current data-collection technology for closing what has come to be known as the data gap. This describes the lack of sufficient decision-quality data to support viable and effective traveler information delivery. Probe data-collection techniques have a superb performance-to-cost ratio, making widespread coverage an economic possibility.

3.4 Telecommunication Technologies

These technologies are ubiquitous in the realm of ATIS. They are incorporated into the data-collection infrastructure to transmit data from the field to centers for processing; they support data exchange between various processing centers; and they support the delivery of the finished information to the travelers and users. Telecommunications technologies can be considered as a mixture of media and protocols that are combined into communication techniques. The media is the physical channel used to carry the data from the origin to the destination. The protocol is the predefined, preagreed content, format, or structure of the data messages to be sent. Let us take a quick look at some popular wireless and wireline communications techniques.

3.4.1   Cellular Wireless

The media of cellular wireless [6] lies at the core of most of the mobile telephone systems in the world today. As the name implies, the technology is based around wireless communications within a well-defined cell with a transceiver at the center. All communications between the transceiver (usually referred to as a cell tower) and mobile handsets within the cell take place using the same radio frequency. Cells are arranged like a blanket across the total area to be covered by the cellular radio system, as shown in Figure 3.8. Care is taken to ensure that the frequency used in each cell is different from those used in the adjacent cells, enabling only three frequencies to cover the entire coverage zone. This is known as frequency reuse and is one of the key features of cellular radio technology as it enables very efficient use of limited radio spectrum. A central communications and computer system is used to link all the cell towers together and link them into the wider telephone network. This system monitors the location of each cell phone that is subscribed to the system and directs calls from the telephone network to the cell tower covering the current location of the phone. Special arrangements are made to hand off the call from one cell tower to another, if the cell phone user is on the move, enabling phone calls to be made seamlessly across cell boundaries.

At the present time most cellular telephone system operators are making use of what is known as second-generation (2G) cellular wireless

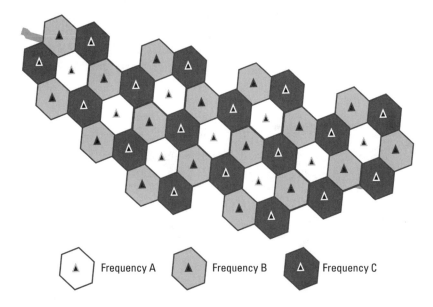

Figure 3.8 Cellular radio.

technologies. First-generation technologies were the original cellular telephone systems that made use of analog communication techniques to deliver voice applications. Second-generation cellular wireless techniques employ digital communication techniques to support both voice and data applications, although the speed at which the data is transmitted is much slower than wireline alternatives. In the near future we expect the emerging third-generation (3G) cellular wireless systems to be widely deployed and to offer high-quality voice and high-speed data applications.

3.4.2 Wireless Application Protocol

In order to make use of current 2G cellular wireless technologies to provide mobile Internet access, the wireless application protocol (WAP) was developed [7]. This is a highly compact and efficient derivative of the HTML technology used to create Web pages on the Internet. It uses the same basic principles and structure of HTML but is specially adapted to the needs of 2G mobile handheld devices, such as cellular telephones and PDAs. Existing Web servers (computers that store and process data for transmission across the Internet) can be linked to an additional WAP server that converts the current Web pages to the WAP format. The information and service available on the Web via home and office computers can then be made available to handheld mobile devices also.

WAP is an enabling technology for mobile access to the Internet. It enables handheld devices, such as cellular telephones to act as mobile Internet terminals. Any Web page from the Web can be accessed from a WAP-enabled device, but most current devices have such limited communication and memory capacity that graphics and memory-intensive pages cannot be read properly. Consequently, many Web site operators have developed modified versions of their Web sites that are specifically configured to WAP devices.

This technology enables a wide range of Internet services to be utilized from a mobile or handheld device, including traveler information services. In the longer term, the evolution of 3G high-speed digital cellular wireless services is likely to supplant WAP as the technology of choice for the delivery of advanced traveler information on the move.

3.4.3 Broadcast Radio and Television

Both of these communications media have become pervasive in many countries around the world, and most people do not even think about how they work. Perhaps this is the true measure of the success of a new technology. When the technology becomes so familiar and easy to own and use that the focus is entirely on the content delivered and the services provided then it has become a mass-market technology.

We have grouped TV and radio together since the broadcast versions work on the same basic principles. A radio signal representing a continuous stable waveform is distorted by utilizing one of a number of different modulation techniques to carry data as well as the original signal. When the combined signal is received by a suitable receiver (radio or TV), the data is removed from the combined signal and converted into either an audio or video presentation to the user. TV and radio originally were broadcast using high-power wireless transmitters. These days, many consumers receive TV and radio transmissions over cable networks and direct-broadcast satellite systems. The next generation TV and radio systems will make use of digital transmission and signal processing techniques to deliver higher quality sound and pictures. These include the emerging high-definition TV (HDTV) and digital audio broadcasting (DAB) initiatives. The transportation community has been making use of radio broadcast techniques for a number of years as a means of advising road users of incidents, delays, and other pertinent information. A radio station dedicated to broadcasting such information is established along with roadside signs equipped with flashing lights or beacons. When there is new information being broadcast, the flashing lights are illuminated directing drivers to tune in to the radio frequency displayed on the sign. In the United States this technique is known as Highway Advisory Radio (HAR). Another more sophisticated technique pioneered by the British Broadcasting Corporation (BBC) in the United Kingdom involves the use of a part of the broadcast radio signal to send data in addition to the original broadcast materials. A suitably equipped radio can decipher the additional data and display it to the user. This technique is known as the Radio Data System (RDS) [8].

3.4.4 Bluetooth

Bluetooth is a short-range digital wireless communication protocol named after a Viking king. The protocol is managed and executed by a microprocessor chip that determines the power required to transmit and receive to and from the nearest Bluetooth-equipped device. Bluetooth will enable short-range communications between devices, such as personal computers, printers, and other peripherals and could form part of a traveler information delivery application in which current information is transferred from a static information point to a handheld device [9].

3.4.5   Copper Wireline Communications

Copper wireline communications include the ubiquitous telephone system that communication engineers refer to as POTS, which stands for plain old telephone system, and advanced wireline communication techniques, such as DSL, ISDN, and T1. These all make use of a single copper wire, or a combination of copper wires to squeeze the maximum amount of data from origin to destination. Potential traveler information applications include the transmission of sensor data from the transportation network to a central processing facility and from the facility to static information delivery devices, such as Internet terminals, kiosks, dynamic message signs, and smart bus stops. The higher capacity techniques may also be suitable for transmitting video images from surveillance devices, such as CCTV cameras.

3.4.6   Fiber-Optic Communications

Where there is a need to transmit very large quantities of data in a short time, most people turn to the use of fiber optics [10]. This technology makes use of the fact that light can be reflected along a long thin fiber or strand of glass for very long distances with little loss of light or signal, through the selection and integration of suitable materials with varying refractive properties. Voice, video, or data can be converted from electrical voltages into light pulses, enabling the data to be transmitted along the fibers at high speed with large data carrying capacities. Multiple telephone calls, data, and video streams can be combined or multiplexed and transmitted along a single fiber optic cable thinner than a human hair.

Due to the large data-carrying capacities and the cost of installing the fiber-optic cables and protective conduits in the ground or on overhead masts, this technology is best suited to long-distance transmission of high quantities of data. This makes it very suitable for carrying trunk communications and broadband signals, such as full-motion video from the transportation network back to a central processing or monitoring facility.

3.5   Data- and Information-Processing Technologies

These technologies support the processing and collation of raw data to produce the desired traveler information output during information-processing and data-fusion activities for ATIS. Data is subsequently stored and retrieved utilizing advanced data-storage and database-management techniques. In this technology area, more than in any of the others addressed in this book, there is huge scope for diving off into great detail. The whole field of database management, data processing, and information science has exploded with activity over the last 10 to 20 years. The advent of low-cost, high-power computing hardware and the maturation of software and application techniques has provided a vast range of enabling technologies and techniques. We cannot do justice to this large body of work while still maintaining our primary focus on ATIS, so we will not attempt to do so. Instead, we will provide the highlights, giving you a flavor of the technology and some insight into the capabilities of the technologies and techniques.

The basic technologies associated with database management and information processing have been around since the late 1930s, but these have been supplemented and enhanced by a vast number of new technologies and developments in hardware, software, and communications. The processing hardware has migrated from the use of valves, diodes, and cathode ray tubes to the use of solid-state devices, such as transistors and integrated circuits. Software has evolved at a remarkable rate from the initial punch cards used to program the early computers to sophisticated computer-aided software engineering tools that take full advantage of the windows, icons, mouse, pointer, and keyboard combination available to support the human-machine interface. In the early days, software programming was carried out in arcane computer assembly language, with every function requiring its own specially developed coding. Today, we have reached the point where interpreters and compilers support the interface between the programmer and the computer and vast libraries of preprepared coding are available for many standard or common functions.

From large isolated processing units with specialist applications, computers have become small, flexible units with the ability to network and communicate with other units. In the communications arena, major strides have been made toward the standardization of computer communications, enabling the development of networks of thousands of computers, such as the Web. These developments have led to a shift in the people perspective as well. In the beginning the end user of a computer system required a team of specialists to develop the hardware and software required to address a specific application, and this is still true to a large extent today. However, the simplification of the human-machine interface, the availability of powerful, easy-to-use tools, and the use of preprepared libraries of software make it possible for the end user to take a much larger role in both the development and management of the application.

To characterize this evolution in hardware, software, and computer communications, we would like to explain briefly the recent development in data warehousing, data mining and visualization, and Internet technologies.

3.5.1 Data Warehousing

This is one of the few subjects on which we choose to provide significant detail, as it has such a significant potential for traveler information system application. One of the most significant features of large traveler information systems is the need to collect and assemble data from multiple sources, and then prepare it for extracting coherent, reliable information. Data warehousing techniques [11] have been developed specifically to address such needs in the most cost-effective and efficient manner, enabling the end users (system operators and managers) to carry out much of the work themselves. There are a large number of proprietary data warehousing techniques and packages available on the market today that support the conversion of large quantities of data into information. Interestingly, most approaches will also support an additional step during which information is converted into knowledge. This goes beyond the simple delivery of the required information to the user and provides decision and analytical support to enable the user to explore and understand the relevance and potential impact of the information, hence unlocking the real value of the information.

Most approaches involve the following steps to the establishment and operation of a data warehouse:

1. *Designing the warehouse:* This involves the selection of the storage and processing hardware, the definition of the interfaces between existing and new computers, and the definition of the software and data handling techniques.

2. *Registering and accessing data sources:* The numerous data sources to be used as feeds into the data warehouse are identified and defined, and access arrangements are established. These would include physical access arrangement and institutional/organizational arrangements, such as agreements to collaborate across different departments or organizations.

3. *Defining data extraction and transformation steps:* The exact methods to be used to extract the data from the numerous data sources and the way in which such data is to be manipulated and

transformed in preparation for entry into the data warehouse are defined in this step.

4. *Populating the warehouse:* In this step the transformed data from multiple sources is entered into the data warehouse.

5. *Automating and managing the process:* Once the data warehouse has been established, a continuous process of improvement and enhancement begins. The users, supported by powerful data manipulation and information-processing tools, make the system more and more automated while managing the overall effectiveness and efficiency of the warehouse.

3.5.2   On-line Analytical Processing

This is commonly referred to as OLAP and is a recent advance in the structured storage and analysis of data in large databases. The type of approach that has emerged and been adopted over the past 15 years or so makes use of so-called relational database structures in which data is allocated to multiple tables with relationships defined between each table. This new approach does not replace the relational one, but adds new capability for faster response times and faster calculation of the results. Instead of the data being held in a large number of tables, it is held in a multidimensional array that can be accessed very quickly. Data consolidation or preaggregation is also carried out for commonly requested analyses, ensuring that response times to queries can be reasonably consistent.

3.5.3   Data Mining and Visualization

This is complementary to OLAP and enables the user by exploring and identifying trends in the database through data extraction, formatting, and graphical representation. These tools provide support for intensive analysis of the data populating the data warehouse, using data search techniques and statistical algorithms to identify patterns, correlations, and trends in the data. This capability supports the conversion of information into knowledge, as discussed above, and is often referred to as knowledge discovery for that reason.

3.5.4   Voice Processing

Voice-processing technologies [12] include synthesis of the human voice, known as text-to-speech technologies, and human voice-recognition technologies.

These technologies can be used to support a range of information delivery options. For example, a text-to-speech application can be used to deliver traveler information to the driver while operating the vehicle, reducing the distraction that could be caused by map and display alternatives. Text-to-speech has also been utilized for a number of years to support interactive voice response systems where information users dial in to a predetermined phone number and press keys on the phone in response to voice prompts to navigate their way around an information store or database. More recently, voice recognition technologies have been added to interactive voice response systems, thus enabling the user to speak navigation commands instead of pressing buttons. Such approaches are converging with Web technologies to enable users to use voice navigation for Web sites, blurring the distinction between interactive voice response and Web applications.

3.5.5 The Internet

The Internet [13] is a global network of interconnected computer networks that provides one of the most valuable information sources and most powerful information delivery mechanisms in existence today. Originally developed in 1969 by the Advanced Research Projects Agency (ARPA) of the U.S. government as the ARPANet, it has since become a publicly accessible, cooperatively operated and managed network. There is a whole range of information and communication technologies that have been developed in support of the efficient and effective operation of the Internet and which fall under the Internet banner. We want to address the following primary ones:

- HTTP;
- WWW;
- Browsers;
- TCP/IP;
- Intranet;
- Extranet;
- HTML;
- Cookie.

Hypertext Transfer Protocol (HTTP) HTTP is the set of rules for exchanging files (text, graphic images, sound, video, and other multimedia files) on the

Web. The basic concept of HTTP is that data files can contain links to other data files, enabling a chain of requests for information.

WWW   From a technical point of view, the Web is defined as all those users and resources on the Internet that make use of HTTP. Tim Berners-Lee, the inventor of the Web, also defines it as follows: "The World Wide Web is the universe of network-accessible information, an embodiment of human knowledge" [14]. In simple terms, the Web is the entire collection of information and knowledge available to you via your Internet connection and your browser.

Browsers   A browser is a software program that enables you to view and interact with Web information sources. A Web browser utilizes HTTP to make requests of Web servers throughout the Internet on behalf of the browser user. The two most prominent commercially available Web browsers in use today are Netscape Navigator® and Microsoft Internet Explorer®.

Transmission Communications Protocol/Internet Protocol (TCP/IP)   TCP/IP is the communication protocol or basic language utilized by the Internet to connect different computer networks and transmit data requests and return data and information. It consists of two separate layers: TCP and IP. TCP manages the assembly of a message or file into smaller packets that are transmitted over the Internet and received by a TCP layer that reassembles the packets into the original message. IP handles the address part of each packet so that it gets to the intended destination on the Internet.

Intranet   This is the term used to describe the use of Internet information and communication technologies within an organization. The technologies are typically used within the context of a local area network that is entirely owned and operated by the organization or enterprise.

Extranet   This is similar to the intranet application described above but incorporates the use of a public shared telecommunications network in addition to the privately owned local area network. This requires the application of security and encryption technologies to keep data secure in the passage through the public elements of the telecommunications network.

Hypertext Markup Language (HTML)   HTML is a software language consisting of symbols or codes. HTML tells Web browsers how to display words and images on the Web page for the user. HTML is generally adhered to by

Microsoft Internet Explorer® and Netscape Navigator®. Both of these, however, implement some features differently and provide nonstandard extensions. The current version of HTML is known as HTML 4.0 and is currently in the process of being replaced with extensible hypertext markup language (XHTML), a new version that allows Web site developers to extend the language to suit their needs. The language is also being developed further to incorporate the use of the markup language to voice devices, such as telephones through the adoption of new voice markup language standards (VoiceXML)

Cookie A cookie is data placed on your local data storage space by a Web server that you have visited. The Web server will use this data the next time you make use of the server to remember your preferences or remember what it provided during your last session. A cookie is a mechanism that allows the server to store its own data about you on your own computer. Cookies are commonly used to customize pages based on your browser type or other information you may have provided to the Web site supported by the server. This technology is particularly relevant to ATIS applications that require the information to be customized to user needs and preferences.

3.6 Display and Delivery Technologies

In this final section we take a brief look at the range of information delivery devices that can be used to deliver traveler information in both static and mobile contexts. These technologies are typically to be found in applications within the information delivery activities associated with ATIS.

3.6.1 Emergency Call Boxes

These are the rugged telephone and emergency communication points found at the roadside, in railway stations, and at some bus stops. It may seem strange to put emergency call boxes in this category, but many current installations have two-way communications capability to and from a transportation-management center. These installations could be viewed as en route points of presence at which the traveler can receive information, provide data, and make requests for service. The exact type of technology used varies from application to application, but most such installations incorporate a power source and communications capabilities. In outdoor applications there is a trend towards the use of solar power cells and batteries to provide the power and wireless communications to support the transmission of data between the installation and a control center. Some basic implementations use a dynamo to generate electricity from the user pulling a handle down to request assistance. At the other extreme, the so-called smart call boxes have the capability for short-range two-way communications with vehicles, enabling the call box to act as a communications hub for the relay of data to and from vehicles on the network.

3.6.2 Kiosks and Smart Bus Stops

In the public transit field, information can be provided by way of information kiosks and so-called smart bus or transit stops. These supplement the conventional marker pole signifying the location of the transit stop with an integrated information terminal and display. The user can access real-time information on the current status of the transit system and obtain an indication of how many minutes will elapse until the next suitable service arrives.

3.6.3 Dynamic Message Signs

In addition to the options described above, traveler information in the form of traffic messages and bulletins can be delivered through the use of information points and displays installed along the highway network. These include variable message signs and information kiosks located at strategic points on the transportation network, such as major decision points for routing, transit stops, stations, shopping centers, and large office complexes. One of the most prevalent applications with respect to road transportation is the variable message sign (also known as changeable, dynamic, or moveable message signs depending on what part of the world you are in). These are typically permanent installations that use a display technology, such as shuttered fiber optics, light emitting diode, magnetic flip disk, or rotating prisms to enable the display to be changed according to circumstances. Some of these signs have fully addressable or bit-mapped displays enabling any combination of text and graphics to be displayed; others have a limited preset number of messages. They can be monochrome or full color. These are used at strategic locations along the highway network to deliver route choice or current traffic condition information.

3.6.4 In-Vehicle Information Systems

These are the systems that are installed in the dashboard on private cars, taxis, commercial vehicles, and transit vehicles. They can be either fully integrated units fitted on the production line or retrofitted units installed as part of an after-market installation. These units consist of a driver information display, an input device, data storage, and processing modules. Figure 3.9 shows a typical unit installed in a private car. The driver can use the input device to specify the type of information required, such as route guidance or traveler information for a specified route or zone of the transportation network. The display may also be supplemented with a voice synthesizer that provides audio information delivery as a means of minimizing driver distraction while driving.

In-vehicle information systems can be autonomous or self-contained in that they have no external communication link to the world outside of the vehicle, or they can be connected via wireless communications. The former approach limits the ability to provide current traveler information to that derived from a historical database probably contained on an onboard data storage device, such as a CD or DVD. The latter support real-time updates and current traveler information and transportation-network conditions delivery to the driver.

While early versions of in-vehicle information systems were specialist units focused on single applications, the advent of wireless mobile Internet services has led to the evolution of more general purpose devices capable of supporting transportation and nontransportation applications.

Figure 3.9 In-vehicle information system.

3.6.5 Personal Information Devices

These are similar in nature to the in-vehicle information systems described above, but these are portable and not linked to the use of a specific vehicle. These include a wide range of devices: cellular telephones, PDAs, and small portable personal computers designed to provide a variety of information services on the move.

3.6.6 In-Home or Office-Based Delivery Systems

The emergence of the Internet and Web has stimulated widespread access to home- and office-based information retrieval and delivery systems that can serve as highly effective delivery conduits for traveler information. Many information service providers offer customized traveler information services that enable the user to specify the nature, timing, coverage, and content to be delivered. Internet communication protocols and standard languages, such as HTML and XML are utilized. These are offered on a subscription basis or free with some sort of advertising subsidy.

3.7 An Operational Concept for ATIS

To complete our treatment of information and communication technologies, we think it is important to provide an overall picture that illustrates how the technologies might be applied in unison to support traveler information system delivery. We call this an operational concept because we have taken features and functions from a number of ATIS to create a theoretical example that illustrates how the technologies can be configured to operate together. Figure 3.10 illustrates our ATIS operational concept, which consists of four information-processing or -management centers and one private-sector information service provider.

Public-Transit-Management Center

The public-transit-management center is operated and managed by the local regional transit authority, operating fixed-route buses and light rail facilities. This center is linked by a combination of wireless and wireline technologies to an automatic vehicle location sensor system and passenger counters that provide an accurate location of each bus and train in the fleet and volumes of passengers boarding and alighting. This data is processed at the transit-management center and compared to a database of scheduled times and locations. The center also has voice and data communication links to every

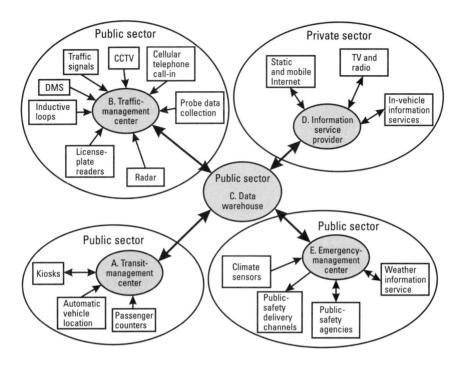

Figure 3.10 ATIS concept.

vehicle to support the transmission of new routes and instructions to operators. The vehicles in the fleet have onboard monitoring systems of vital components, such as oil, temperature, fluid levels, brake wear, and tire condition
in addition to passenger counting equipment. Data from these devices is also
transferred back to the transit-management center for fleet-management and
passenger-information purposes. The center is also connected to a range of
information kiosks at strategic locations, such as bus stations, rail stations,
and other public places, enabling current travel information to be delivered
to these locations.

Traffic-Management Center

The traffic-management center is operated and managed by a public-sector
transportation agency responsible for traffic control and management on
both high-speed limited access highways and urban surface streets and arterials. The center is linked by wireline and wireless communications facilities to
a network of traffic sensors. These include CCTV cameras, inductive loops
for vehicle speed and traffic volume measurements, and license-plate readers

and toll tag probe data readers for link travel time estimation. A cellular telephone call-in dispatching system enables travelers to report traffic conditions and incidents to the traffic-management center. The center also controls a range of roadside dynamic message signs that enable current travel and traffic information messages to be displayed to drivers. On urban surface streets, the center is connected to a network of traffic signal controllers, inductive loop vehicle detectors, and traffic signals that collect data on current traffic conditions and control traffic flow through the adjustment of signal timings in line with traffic conditions.

Data Warehouse

The data warehouse is a centralized data-processing facility owned by the public-sector agencies on a joint basis and operated by a private-sector contractor on behalf of the public agencies. This data-processing facility collects the data available from each of the public-sector management centers—transit, traffic, and emergency—collates, cleans, and verifies the data, and develops a single data repository that can be accessed by all three public agency centers and the private-sector information service provider.

Information Service Provider

The information service provider is operated and managed by a private-sector enterprise in the business of collecting and processing traveler information and travel-related data, developing information services and products, and delivering them through multiple delivery channels to subscribers. The information service provider is linked by wireless or wireline communication facilities to the central data warehouse and receives processed data and video from the warehouse. The information service provider also uses the communication link to send fused and value-added data back to the warehouse.

The information service provider also has a range of information delivery channels that have been developed and established, including fixed and mobile Internet access, in-vehicle information systems, an interactive voice response system, and TV and radio broadcasts.

Taking a look at how this "system of systems" operates in a routine context, the current traffic conditions and transit operational status data is collected and delivered to the data warehouse for storage and processing. In a nonemergency situation, the emergency-management center would be inactive and may just process this data to develop historical traffic pattern and travel trend information. The information service provider draws a preagreed data set from the data warehouse and carries out additional information processing in order to develop the data streams required to feed the multiple information delivery channels available. Video images from the traffic CCTV cameras are packaged and delivered to participating local and regional TV stations, and voice travel bulletins are developed and delivered to local radio stations. Data is also converted into free and subscription information products and services for delivery to static Internet access points for pretrip and en route kiosk applications, as well as to mobile handheld devices, such as WAP cellular telephones, pagers, and PDAs for en route applications. Data or video is packaged and provided to other content providers who will integrate information into their services. The information stream to the traveler has a full range of traveler information including current travel conditions, alternative routes and modes, and relative times and costs for the journey. Information can be customized for the individual traveler as required.

Emergency-Management Center

The emergency-management center is operated and managed by the local public safety agency responsible for emergency planning and management for hurricanes and other natural and man-made disasters and special events. This center is linked by specially protected and redundant wireline and wireless communication facilities to the regional and national weather information service, police, fire, ambulance, and other relevant government agencies for emergency management, including the three transportation-management centers described here. It is also linked by wireless and wireline to field operatives and climate sensors, such as flood gauges, wind gauges, and temperature gauges. This center also has links to a range of public-safety information delivery channels, such as radio and television to be used to broadcast emergency messages and instructions, such as evacuation notices.

In the event of an emergency situation, such as a hurricane, forest fire, major flooding, or terrorist attack, the emergency-management center is activated, adding new data sources and needs for information delivery. The emergency-management center develops and delivers evacuation plans and assumptions to the central data warehouse for relay to the traffic- and transit-management centers. At the traffic-management center these plans and assumptions are used to develop a range of traffic-management strategies for supporting the emergency-management center's needs and underwriting assumptions made regarding evacuation times from zones. These may include plans to reconfigure road use to accommodate one-way, tidal-flow operations away from the impending disaster, the use of diversion routes, and the removal of tolls and other road use charges. Similarly, at the transit-management center, plans are developed to deploy the transit fleet to optimize the efficiency and effectiveness of the evacuation plan. The communication links between the three centers and the data warehouse are used to share proposed plans and develop a coordinated, collaborative approach to travel management during the special event.

The information service provider supports the emergency-management activities by developing and delivering real-time information and advice to residents and travelers using multiple-channel information delivery systems. The information is also fed back through the central data warehouse to be relayed to public zone-specific information delivery systems, such as TV, radio, and reverse 911 systems. In the United States, these reverse 911 systems are being used to push information and instructions to residents by automatically dialing all telephone numbers within a predefined zone and delivering an automated advisory or message. The data and information delivered is also stored in the data warehouse for later evaluation of performance and the development of emergency and traffic-management strategies.

We have just provided a fictitious example that makes use of many of the ATIS technologies we described earlier. It makes a great deal of assumptions regarding who does what in the overall system. For example, we assume that the private-sector information service provider would own and operate the Internet channels in the information delivery system, with the public-sector traffic-management center owning and operating the roadside dynamic message signs. We also assume that all system operators share data to and from a central data warehouse facility that acts as the bridge to the private sector. The decision on who owns, operates, and manages the various subsystems is not a trivial one and requires careful consideration of objectives and capabilities. As will be discussed in Chapter 6, a wide range of options and alternatives need to be considered and evaluated.

3.8   Summary

In this chapter we have attempted to provide a very broad overview of the range of information and communication technologies that can be applied to ATIS. This is intended to give you an overall awareness of the technologies that are available, and a general idea of how they work and what they can do. The main impression that we want to leave with you is that the range of technology options is very wide and is increasing almost on a daily basis. Progress in the development of new telecommunication techniques and the use of the Internet seems to be accelerating, with new technologies emerging and new combinations of existing technologies being developed and implemented. On this latter point, the convergence of technologies that were originally considered as separate entities is one of the most striking trends today. Taking the Web and interactive voice response technologies as examples; these at first seem to be quite separate technologies with Web technologies, such as HTML and XML, addressing the delivery of information via the standard computer setup of display, keyboard, and processor, while the interactive voice response technologies support the delivery of information via a standard telephone. Recent advances, however, in the extension of HTML and XML to include VoiceXML make it possible to converge a traveler information Web site with an interactive voice response system. The interactive voice response system simply becomes a way of navigating the single Web site by using speech recognition and speech synthesis technologies. We expect to see many more examples of such technology convergence over the next few years.

To conclude the chapter and as a way of illustrating the application of the various technologies within the framework of an ATIS, we have provided an operational concept. This describes an imaginary implementation of information and communication technologies to transportation operations and management. We use this to show how the range of technologies might work in harmony to deliver traveler information while integrating with other transportation- and emergency-management functions in a region.

We are aware that our high-level approach to the identification and description of information and telecommunications technologies may have left you with more questions than answers and may have created a thirst for more knowledge and information. You may pursue the references below if you are interested in learning more about the technologies we have described in this chapter. We make no apologies for this. If we have raised your awareness of information and communication technologies and influenced your thinking on the subject, then we have done the job we set out to do in this chapter.

References

[1]   EIS RTMS, http://www.rtms-by-eis.com/index.html.

[2]   Texas Transportation Institute, Sensor Testbed Research, http://transops.tamu.edu/fraRsch.htm.

[3]   U.S. Coastguard Navigation Center, http://www.navcen.uscg.gov/loran/default.htm.

[4]   "Detection Technology for IVHS Volume I: Final Report," Research and Development, Turner-Fairbank Highway Research Center, McLean, Virginia, December 1996; also at http://www.itsdocs.fhwa.dot.gov/jpodocs/repts_te/4rs01!.pdf.

[5]   Computer Recognition Systems, http://www.crs-its.com/; and Pearpoint Inc., http://www.pipstechnology.com/about.html.

[6]   Rappaport, T. S., *Wireless Communications: Principles and Practice,* Englewood Cliffs, NJ: Prentice Hall, 2001.

[7]   Foo, S. M., C. Hoover, and W. M. Lee, *Dynamic WAP Application Development,* Greenwich, CT: Manning Publishers, 2001.

[8]   Kopitz, D., and B. Marks, *RDS: The Radio Data System,* Norwood, MA: Artech House, 1998.

[9]   Miller, B. A., and C. Bisdikian, *Bluetooth Revealed: The Insider's Guide to an Open Specification for Global Wireless Communications,* September 25, 2000.

[10]   Hecht, J., *Understanding Fiber Optics,* June 14, 2001.

[11]   Ponniah, P., *Data Warehousing Fundamentals: A Comprehensive Guide for IT Professionals,* New York: Wiley Interscience, 2001.

[12]   Abbot, K. R., *Voice Enabling Web Applications: VoiceXML and Beyond,* Berkeley, CA: APress, 2001.

[13]   Levine, J. R., C. Baroudi, and M. L. Young, *The Internet for Dummies,* New York: John Wiley & Sons, 1999.

[14]   Worldwide Web Consortium, http://www.w3.org/www/.

4

Public Objectives, Private Enterprise

4.1 Learning Objectives

After you have read this chapter you should be able to do the following:

- Define a range of public-sector organization roles and types with respect to traveler information systems;
- List a candidate set of public agency transportation objectives;
- Define a range of private-sector organization roles and types with respect to traveler information systems;
- List a candidate set of private enterprise traveler information system objectives;
- Distinguish real objectives from metaobjectives;
- Define at an outline level, the potential areas for public-private collaboration in ATIS activities.

4.2 Introduction

This chapter will explore the range of public- and private-sector organizations and enterprises that may be involved in the development, establishment, and operation of an advanced traveler information system. We want to ensure that you understand the full range of entities that could be involved and provide you with an insight into their motivations for being involved. This should help you to decide who should be involved and who should be consulted when you are developing your plans for advanced traveler information systems.

The world of ITS presents many opportunities for collaboration, cooperation, and mutually beneficial actions involving both the public and private sectors. These include the more conventional client-vendor relationships and a range of new mechanisms involving some degree of partnership or sharing. We call these new mechanisms, but, of course, many of them have been applied successfully for a number of years in other application fields and are just new to transportation and ITS. A good example would be the so-called shared resources approach to the planning, development, and implementation of fiber-optic wireline communication systems where the public sector allows private access to right of way on public roads in return for the use of some of the communication capacity installed by the private sector.

ATIS present a range of feasible partnering possibilities where the public sector can enjoy similar leverage to the private sector's desire to exploit business opportunities associated with the application of information and communication technologies to transportation. Like most partnership possibilities, those associated with traveler information systems can only be truly effective and of sustained value to all partners if the needs, issues, problems, and objectives of all partners are fully understood. Indeed, each partner should have a good grasp of their organization's respective objectives and requirements before opening a dialog with other partners. As with most things in life, getting what you want usually requires asking for it. How can you effectively communicate what you want if you do not really know yourself? How can you represent your organization around the bargaining table if you do not have a clearly defined, agreed-upon set of objectives, constraints and other parameters defining an acceptable bargain?

It is, of course, very easy for us to say this and harder for you to address the questions and follow our advice. In our experience a circular process develops in that you need to know something about your range of choices and options before you can start to effectively answer the questions regarding what you want. In order to provide some support in this respect we have conducted a thorough review of all the needs, issues, problems, and objectives that we have encountered in the course of our traveler information consulting work. We have compiled a range of objectives based on this review and present them in the next section for your information and review. Note that we have selected each of the needs, issues, problems, and objectives so that they all have a potential solution within advanced traveler information systems. This is our way of breaking the circle and getting you started on the road to a good clear understanding of your needs relative to what ATIS can achieve. You should also note that it is perfectly acceptable to define an initial set of objectives based on the information presented here, then come back and revise the objectives in the light of additional information you have gathered about ATIS capabilities or private-sector service offerings.

The identification, exploration, and confirmation of objectives in the form of a sound requirements analysis should be a key first step in the effective planning and development of any ITS (or any system for that matter). We have conducted numerous requirements analysis exercises in the course of our consulting work on ITS and on ATIS in particular. These have involved the definition and establishment of a stakeholder group comprised of the main players in the transportation context within a region.

These are people and organizations that are either directly involved in planning, development, funding, and implementation and those who will be affected or impacted by the proposals. From these various activities with a range of stakeholder groups, we have distilled the following list of potential public-sector objectives and motivations when considering the deployment of an advanced traveler information system. If you are in the public sector, take a look at them and see if you can relate them to your agency. If you are in the private sector, consider this to be essential information on your potential partners.

The best examples of the application of information and communication technologies to the delivery of traveler information start with a basic definition of what has to be achieved. This may seem obvious, but many applications of advanced technologies fail to achieve the desired effects and do not meet expectations for the simple reason that nobody defined and agreed upon objectives at the beginning. When there are several partners involved, it becomes even more important to define and agree upon needs and objectives and to understand the motivation of all partners from the outset of the initiative.

In this chapter, we explore the nature of public transportation agency objectives for traveler information systems. We contrast these against the needs and objectives of the private sector, highlighting common ground and significant differences. Our intention is to provide you with a solid list of objectives that you may wish to use as a starting point for the definition of your needs. We also hope to shed sufficient light on both sides of the potential public-private partnership fence to provide the basis for effective and efficient public-private partnership identification, establishment, and maintenance.

This is one of the most important parts of our book. Earlier chapters have prepared the ground by providing you with information to allow you to understand this chapter and utilize it effectively. Here we will explore the nature of public- and private-sector objectives associated with the planning, development, promotion, and implementation of traveler information systems. Starting with the public sector, we will explore a range of potential objectives, which are based on the authors' collective experience in working with public-sector agencies to define traveler information system objectives. We will explore the relationships between objectives and actions in traveler information systems and the wider mission of public-sector agencies.

Before moving on to the private-sector perspective on traveler information system objectives, we will take some time to explore the new situation that public agencies find themselves in when they attempt to have their objectives addressed by other people's money. This addresses the issues and challenges associated with partnership, including the need for influence and advocacy rather than direction and control.

Then we will take a close look at the private sector's objectives in being involved in the traveler information systems business, by exploring motivations, perspectives, and needs from a commercial organization's perspective. Finally, we will explore the differences and the commonalities between the public- and private-sector perspectives. The language we have used to describe the two perspectives—public objectives, private enterprise—suggests that there may well be cultural differences and significant variation in needs and objectives.

This chapter should provide you with a detailed overview of the range of objectives that public agencies may satisfy through involvement in a traveler information system, while illustrating the need for such work to be objectives driven. It should also give you a detailed understanding of the new paradigm that public agencies must manage if public objectives are to be effectively addressed through partnering rather than autonomous initiatives. We hope that this chapter will give you valuable insight into the motivations and objectives of private-sector organizations engaged in the traveler information industry and the cultural differences between public and private perspectives. With this knowledge you should be able to identify and support more effective partnership approaches to ATIS.

4.3 Public Objectives

In our exploration of the objectives that public agencies might define with regard to traveler information systems, we will identify and define the type of organization that could represent the public sector, define some typical roles within the organizations, and then develop a list of candidate objectives.

4.3.1 Organization Types and Roles

When we started to put together the initial list of objectives, we thought it best to look at the world of traveler information from the perspectives of a range of typical roles that we have identified in public and private organizations. This helped us to make sure that the ensuing list was firmly rooted in reality and as complete as we could make it. As we were doing this, it became clear that if we were going to be really helpful to you as the reader and investor of time in reading this material, we should not only provide the list of objectives, but also a description of the typical roles we used to guide the development of the list.

Therefore, we start off each exploration of objectives by taking some time to explain the types of organization we had in mind and the range of typical roles within those organizations that we have encountered in the course of our traveler information systems development work. Sections 4.3.1.1 to 4.3.1.11 list the types of organizations and professionals we have in mind when we think about the public-sector part of traveler information.

4.3.1.1 Central, Federal, and National Government Departments of Transportation

These are the national public agencies responsible for planning, implementing, and operating the transportation network. In most countries, such agencies have responsibility for national coordination of transportation policy and national expenditure as well as an operational role with respect to nationwide infrastructure, such as the network of major highways.

4.3.1.2 State and Local Departments of Transportation

These are regional and local public agencies with specific responsibilities for the planning and coordination of transportation within a defined region or state within a country. Such agencies may apply nationally developed policies and expenditure acting as agents of the central or federal agencies described above. They may also develop and implement their own local transportation policies and transportation investment programs.

4.3.1.3 Public Transit Authorities

These could be considered as special types of state and local public agencies, with specific focus on public transit or transportation, such as bus, rail, and light rapid transit systems within a region. In some countries and regions, the public sector is only responsible for policy setting and overall coordination, with the role of operating the transit system assumed by the private sector.

4.3.1.4 Railway Agencies

These organizations are responsible for intercity rail travel and the development and ongoing management of the national rail system. Here again, in some countries, the role of the public sector has been confined to policy, coordination, and perhaps operation and management of the fixed infrastructure, assigning the role of operating the rail services and vehicles to the private sector.

4.3.1.5 Airport Operating Agencies

These can be city-specific, regional, or national organizations that are responsible for developing and operating the air terminal and passenger handling facilities at airports within their jurisdiction. Such agencies typically have responsibility for road-traffic management and parking management in and around the airport facilities.

4.3.1.6 Maritime Administrations

This group included those organizations responsible for the development and operation of sea-going passenger transport, such as ferries. In some cases these responsibilities are assumed by the local department of transportation; in others a port authority takes responsibility for this and other management functions related to the movement of passengers and goods in and out of the port.

Within these organizations, we have defined a range of typical roles, based on many we have encountered in the course of our traveler information work.

4.3.1.7 Operations Professionals

This group includes roles like traffic engineer, freeway manager, toll road operator/manager, transit operations manager, and landside operations manager. As operational-management professionals, this group places a heavy emphasis on the use of information and communication technologies to improve the efficiency of the operation of their transportation facilities and provide higher levels of customer service. In most cases, members of this group try to maximize the available capacity, or supply of transportation, ensuring that this matches current demand.

4.3.1.8   Transportation Engineering and Planning Professionals

Transportation engineering and planning tends to encompass several modes of transportation. Professionals engaged in the field try to identify demand for transportation and understand the fundamental mechanisms that drive that demand. They then try to forecast future demand on the basis of this understanding and develop longer range plans to provide the right transportation resources in the right place at the right time. Engineering tends to focus on the design and delivery of the transportation resources required, while planning focuses on the identification of needs and the understanding of transportation demand mechanisms, such as land use.

4.3.1.9   Procurement Specialist/Contracts Professionals

This group of roles involves the development of scopes of service, statements of work, and procurement and bid documents that specify what the public-sector agency wishes to procure or acquire and under what terms and conditions they are prepared to enter into an agreement to do so. People in this group have a duty to ensure that public funds are being applied in an optimum manner while preserving social equity and fostering or even promoting an open, fair, competitive market for goods and services.

4.3.1.10   Public Information, Marketing, Customer Service Professionals

This group emphasizes the development and delivery of information to the public on services being provided and value being delivered. This includes the communication of the rationale for taking specific actions and choosing certain approaches. Information on how to gain access to services and the potential value and benefits of the services delivered are also part of the focus for this group.

4.3.1.11   Political Professionals

This group includes government officials, federal and local politicians, board members, council members, and other leadership positions elected by democratic means. The aims of this group are to deliver the actions and values promised and to serve the electorate and thus maintain their status as effective, elected officials. Consequently, this group has a strong focus on visible, tangible impacts, benefits, and values.

4.3.2   Candidate Public-Agency Traveler Information Objectives

Public-sector agency objectives with respect to traveler information and transportation typically revolve around three primary categories:

1. Environmental;
2. Economic;
3. Social.

For further illustration, [1, 2] provide an overview of current central government policy objectives in the United States and Australia, in the form of high-level policy statements and mission statements. These three categories could be considered as the intrinsic values that public agencies pursue through their interest and activities in ATIS development and operation. We can also define a range of candidate objectives that support these three major categories. These can be considered as instrumental values that are pursued in support of the intrinsic ones [3] as follows.

4.3.2.1 Obtaining the Best Value for Public Funding

Always high on the list of public-sector objectives for any project, program, or initiative, this addresses the public organization need for tangible efficiency and cost-justification. Most public-sector agencies have a strong focus on a range of value objectives that can usually be traced back to three things: saving time, saving lives, and saving money. Public funds are typically highly constrained, requiring very good cost justifications to be prepared before a project can receive any significant degree of public funding. Public agencies are strongly motivated to demonstrate clear cost-effectiveness and wise investment decisions. A great way of approaching this is to make sure that any proposal for funding has been subjected to a thorough and objective analysis of the life-cycle costs incurred and the total benefits delivered. Costs are derived from nationally accepted values based on current market conditions and experiences of other previous implementers and procurers. Benefits are determined from past experience and from simulation modeling of the proposed implementations.

In addition to the determination of the intrinsic value of a particular implementation, the public sector is also interested in ascertaining the perceived value or likely public impact of the proposals. Such perceived value has a significant impact on the general public's opinion of the usefulness and effectiveness of public agency actions on their behalf.

4.3.2.2 Maximizing the Use of Legacy Systems and Sunk Investment

Building on the need for tangible demonstrable benefits and cost-justification, it would be very difficult for a public agency to build such a case, move forward with deployment, then reverse its position a short time after deployment, throw everything away, and start again. Technology is dynamic and there are situations where greater benefits can be attained at lower cost by scrapping legacy systems and replacing them entirely. Replacing a legacy system before the end of its design life, however, is likely to cause significant angst within a public-sector organization. The early adopters who championed the legacy system may still work within the agency (often in very senior, influential positions), and reputations and future careers may be built around the legacy system.

As an aside, we have just made an important assumption that is worthy of a few words. We talked about the difficulty of replacing a legacy system before the end of its design life. This assumes that there is a clear, agreed-upon understanding of the intended design life of the system. Surprisingly, there are many examples of ITS implementations around the world where no intended design life has been established, making it difficult to determine whole life costs, benefits, and the timing of the break-even point where benefits outweigh costs.

4.3.2.3 Leveraging the Efforts of Others

In the interest of avoiding waste and duplication, public agencies are extremely interested in making full use of the efforts of other public and private organizations within their region or jurisdiction. The prime motivation here is to avoid the reinvention of the wheel through understanding what others are planning or doing and building this information into agency plans and initiatives. An important secondary motivation is to go beyond avoiding overlap to seek synergy in the form of new revenue streams that can be harnessed to improve the quality of public infrastructure and services. One aspect of this need to leverage the efforts of others is the strong desire to ensure that traveler information systems deployment are fully compatible and integrated with other ITS deployments and applications within the jurisdiction of the public agency. In fact, the development of the National Architecture for ITS in the United States and the subsequent federal rule-making have made this a regulated requirement for U.S. deployments of ITS technologies, products, or services receiving federal funding. The National Architecture for ITS is a high-level conceptual framework that identifies and defines all the major subsystems and their communications links required to satisfy a predefined range of objectives. Derived from a set of user needs that were identified, defined, explored, and confirmed through formal requirements analysis, the National Architecture for ITS paints a future big picture for the deployment and operation of information and communication technologies applied to transportation in the United States. We will discuss this in more detail when we examine the big picture context for ATIS in Chapter 7.

4.3.2.4 Addressing as Many Organizational Objectives as Possible

Another way to measure value for money in a more subjective fashion is to evaluate the proposal against the overall stated objectives of the public agency. Proposals that address a larger number of the agency's stated goals, or have a better fit with overall mission and directions would fare better under this type of analysis. This approach would be more useful if the relative impact of achieving each of the objectives was also considered, as this would provide a better measure of effectiveness.

4.3.2.5 Investing Public Money in Opportunities Appropriate for Common Funding

A tough issue for public-sector organizations rests in the appropriate application of public funds. With a public position that requires the fair and equitable use of funds that have been raised through central and local taxation, the public sector requires that any investment of public funds must satisfy criteria that measure this important facet. This poses a particular problem for public-private partnerships, as the public sector has to be careful not to subsidize or underwrite a private venture that would give the private partner a monopoly or distort the evenness or fairness of the local market. The public sector must also define and adopt an appropriate posture with regard to risk and liability to ensure that public funds are subjected to the appropriate exposure. Appropriate uses of public funds may also be defined and bounded by procurement and contract regulations or by policy statements that have been adopted by the agency. These may extend to the definition of a formal procurement process for goods and services that must be adhered to for all procurements, irrespective of nature.

4.3.2.6 Leveraging the Motivation and Desire of the Private Sector

In the same manner as they attempt to leverage the efforts and activities of other public agencies, the public sector is very keen to incorporate the efforts of the private sector. There is a growing recognition that many of the activities in the chain that leads from data collection to information delivery are best suited for the private sector. The convergence of public objectives with private enterprise in information delivery, in particular, requires that public agencies take this posture.

This is an important factor in the quest to make public expenditure as efficient as possible by avoiding duplication and overlap with current private-sector activities. For example, if there happens to be an extensive privately developed and operated information delivery mechanism that fully meets public needs, then the public sector should give serious consideration to making use of the existing mechanism instead of developing a parallel one that may in fact compete. In some countries, the public sector is expressly forbidden to use public funding to compete directly with the private sector. We will return to the nature of the traveler information supply chain and public agencys' need to adopt certain management techniques and approaches in Chapter 6 when we discuss ATIS business models and approaches.

4.3.2.7 Traveler Behavior Change

We think this is one of the most fundamental parameters that public agencies should define and agree upon prior to significant involvement in the development and delivery of traveler information systems. When you think about it, the desired change in traveler behavior can be closely linked to many of the public-sector transportation policy objectives and goals possible. Behavioral changes, such as adopting a different route to work, departing earlier or later, and making use of alternative modes of transportation can, if made in significant numbers, have a great effect on transportation-network conditions. For example, a decision by 10% by those travelers who normally travel at the height of the morning commute to revise their departure time to one hour ahead of or behind the peak could have a substantial effect on reducing recurring peak period congestion. It would seem that the precise set of needs, issues, problems, and objectives to be addressed by a proposed traveler information system varies from one region to another based on current transportation-network conditions and other parameters defining the region's starting condition. Consequently, the desired traveler behavior changes will also vary to some extent from one region to another.

As we will discuss in Chapter 7, ATIS could also be viewed as a softer way to impose transportation management and control. Through the provision of accurate, timely, comprehensive, and reliable traveler information that forms the basis of traveled decision making, travel behavior and demand can be altered and influenced. Another important effect of this is increased traveler satisfaction and improved travel experience, which can be supported through the provision of information that leads to more effective use of the transportation infrastructure. Stepping back from the whole subject for a minute and considering why public agencies are involved in the development and operation of advanced traveler information systems, we believe that this objective will be near the top of the list.

4.3.2.8   Jurisdiction-, Mode-, or Corridor-Focused Activities

Because of the way they do business, most local and central government agencies have a focus on planning and implementing ITS on a project, corridor, or jurisdictional level. This is mainly because this is how government agencies maintain a focus on getting things done within their domain or authority. This works extremely well for asphalt-, concrete-, and steel-type projects where high value, long-term transportation infrastructure is being planned and implemented, but it may not be the best way to approach information and communication technology initiatives with private-sector partners. Many public agencies, however, are organized in terms of transportation policies, programs, and projects to take a narrow focus on specific parts of the transportation network. Needs, issues, problems, and objectives that are specific to the parts of the networks or corridors involved should be identified and confirmed and then incorporated into ATIS design and development. It may be necessary to widen requirements in order to cover a larger geographic area in addition to the target one, to make the proposition commercially viable or attractive to prospective private-sector partners.

4.3.2.9   Promotion of a Particular Travel Mode

In order to satisfy policy goals, a public-sector agency may desire to promote the use of a particular mode of transportation or certain routes. For example, in order to address goals to reduce emissions or traffic congestion, it may be the goal to promote the systemwide use of public transport or perhaps express bus services on certain commuter routes. This leads into the realm of transportation demand management, where policy makers, transportation planners, and operators attempt to influence the demand for travel as a means to match supply and demand. This type of objective requires careful treatment in the development of a partnership with the private sector, as information designed to influence travelers in a specific direction may not have the same commercial appeal as information that simply provides a complete picture of current conditions and options, leaving the traveler to decide which is the best option to take.

4.3.2.10   Reducing Intermodal Travel Times

Another public objective could be to enhance the possibilities for intermodal travel by reducing the time taken to transfer from one mode of travel to another. This would entail a combination of physical measures, such as the design and implementation of modal transfer points like bus stations, combined bus and train stations, and park-and-ride facilities, and the provision of traveler information on how to make the transfer, how long it will take, and how much it will cost. It should also entail the synchronization and coordination of transportation services to support seamless payment and minimum wait times at transfer points. ATIS have a potentially large role to play in the promotion of intermodal travel and in optimizing the effectiveness and efficiency of the transfer. They lend themselves to the delivery of information on a range of options for trip chains and combinations of travel modes. ATIS can also be evolved into highly sophisticated decision-support facilities that provide the information and analyses in support of more complicated evaluation of alternative and more sophisticated decision making on the part of the traveler.

4.3.2.11 Improving Quality of Service

Public agencies have a duty to ensure that the quality of transportation services is as high as it can be. They measure the quality of transportation in different ways, according to the mode of travel. Here are a few sample quality or performance measures:

- Reduced traveler stress, increased user satisfaction;
- Improved safety;
- Reduced travel times and congestion;
- Reduced travel costs;
- Reduced intermodal transfer times;
- Increased service or capacity availability;
- Increased utilization of transportation facilities;
- Increased travel-time reliability.

The last performance measure—increased reliability of journey time—is of particular interest. Recent research work in the United States [4] has identified that the likely principal benefit of ATIS to travelers will be the increased reliability in trip travel times and the corresponding decrease in slack time that must be allocated to a given journey. This ability to reduce or eliminate the slack or buffer time that travelers build into their journey plans has the potential to unlock considerable value. This is especially true when you consider the effects of just-in-time delivery on freight transportation. At the moment, the trucking industry ensures that goods are delivered on time (avoiding hefty financial penalties) by arriving ahead of schedule and tying up the driver and the truck by waiting at the destination until the appointed delivery time. In more general terms, the ability to accurately predict travel times through the transportation network and, hence, underwrite the reliability of journey times is a major indicator of how well we can manage and control our transportation networks. Another aspect related to service improvement is the collection and analysis of data that can be used to measure and monitor the performance of the transportation network. This all reminds us of the fundamentals of managing anything: First you have to understand how it works; then you have to be able to monitor and measure its performance; then you have to be able to manage it.

4.3.2.12 Maintaining an Appropriate Level of Control and Management Direction

In order to have the capability to ensure that transportation policy objectives are being successfully addressed, public agencies must insist on a certain degree of control and a full definition of management and control arrangements for any implementation or initiative involving public funds. The public sector wants to ensure that the public good is safeguarded and transportation objectives are achieved. In a free market context there may be multiple traveler information service providers providing traveler information; this may have an uncoordinated, unpredictable effect on travel behavior and the transportation network. The public-sector agencies would wish to ensure that they had sufficient management and regulatory control to deal with this type of issue. This also relates to the attainment of what we would call *community optima* with regard to the operation and management of the transportation network. We see the public agencies acting on behalf of and in the best interests of the entire community to ensure that benefits are achieved and available in a fair and equitable manner. This may not always be in agreement with the objectives of the private sector. If the private sector has adopted a business model that revolves around the sale of traveler information services and products to subscribers, then there may be a tendency to seek the personal optimum that delivers the highest value for the individual paying the subscription, but may result in suboptimization of the overall transportation operation. For example, revealing details of a shorter, cheaper, or faster route through the transportation network to only those who subscribe may not allow the whole network and all journeys to be optimized.

Table 4.1 summarizes the list of public-sector objectives identified and provides a tentative mapping of the objectives to the roles as previously defined.

Table 4.1
Summary of Public-Sector Objectives

Role	Obtaining the best value for public funding	Maximizing the use of legacy systems and sunk investment	Leveraging the efforts of others	Addressing as many organizational objectives as possible	Investing public money in opportunities appropriate for common funding	Leveraging the motivation and desire of the private sector	Traveler behavior change	Jurisdiction-, mode-, or corridor-focused activities	Promotion of a particular travel mode	Reducing intermodal travel times	Improving quality of service	Maintaining an appropriate level of control and management direction
Operations professionals	●	●	●		●			●	●			●
Transportation-engineering and -planning professionals	●	●	●	●	●	●	●	●	●	●	●	●
Procurement specialist/contracts professionals	●	●	●	●	●	●					●	●
Public-information, marketing, customer-service professionals	●	●	●	●	●	●	●	●	●	●	●	●
Political professionals	●	●	●	●	●	●	●	●			●	●

4.4 Private Enterprise

Having spent the first half of this chapter talking about the needs, issues, problems, and objectives of the public sector, we now switch to a focus on the private-sector organizations and enterprises that could potentially be involved in the world of traveler information.

When we take a look at the delivery of traveler information from the private-sector perspective, we see the world through a very different lens. Motivations, operating context, and overall goals can be significantly different from those of the public sector.

4.4.1 Organization Types and Roles

The private sector has the potential to bring a much more diverse set of organizations and enterprises to participate in the world of traveler information. The traveler information value or supply-chain features numerous points where the combination of an information or content stream and a suitable delivery mechanism or device could be a potent engine for economic gain. Traveler information is one of a number of such streams of information that have commercial value and can form the basis for a private enterprise. Others include weather, stock quotes, sports scores, and yellow pages information. Since traveler information could be offered as one of a bundle of such services, a wide range of private-sector enterprises may be involved in traveler information. These include the following nonmutually exclusive categories.

4.4.1.1 Traveler Information Service Providers

These include the private-sector enterprises that specialize in the development and delivery of information services and products. They can deliver information directly to the consumer or act in a wholesale manner by delivering information to other service providers for incorporation in wider distribution channels and information service and product bundles. Some may have their own data-collection and information-processing infrastructure, while others may rely on public data sources.

4.4.1.2 Personal Information Service Providers

There are many other information service providers that develop and deliver a range of information services and products that address sports scores, stock prices, weather, yellow pages, and news services as well as traveler information. Some of these service providers deliver customized information packages tailored to the needs and preferences of specific consumers or consumer groups.

4.4.1.3 Telecommunications Service Providers

This group includes the myriad of public- and private-sector organizations that own and operate the wireline and wireless telecommunications infrastructure used to deliver much of the traveler information to the consumer and transport the raw data from the various field locations to and between information-processing facilities.

4.4.1.‡ Automotive Electronics Manufacturers

These organizations develop and supply the electronics and the information/communications technology-based products that automotive manufacturers install in their vehicles, including in-vehicle entertainment and information systems.

4.4.1.5 Automotive Manufacturers

This group consists of the manufacturers and suppliers of the cars, trucks, and buses that operate on our highway infrastructure. While these organizations are primarily equipped and motivated to sell vehicles, there is an increasing trend towards the manufacture and delivery of information and communication technology options built on the basic vehicle platform. In the United States this has come to be known as telematics.

4.4.1.6 Advertisers

These organizations could be any one that has a product or service to sell, which requires prospective buyers to be informed about the features and benefits.

Within each of these enterprises we have been able to define a range of typical roles and responsibilities just as we did for the public-sector earlier in this chapter. These are based on our experiences in developing public-private partnerships with respect to traveler information systems and our operational experiences on selected projects in the United States. The list is not intended to be comprehensive, but indicative of the types of roles you might find in a private enterprise engaged in traveler information.

4.4.1.7 Business Management

This role includes the establishment and agreement of the overall business direction of the enterprise, usually through the development and communication of a formal business plan. This would be agreed upon either with in-company enterprise sponsors or with external venture capital providers or other sources of external finance.

4.4.1.8 Operational Management

This is the day-to-day management and control of the activities that constitute the business, value, or delivery chain for the enterprise. In the case of a traveler information service provider, this may include the operation of a data-collection and information-processing system linked to a multichannel information delivery system.

4.4.1.9 Marketing and Distribution

Identifying the target markets or market segment for the goods or services to be supplied by the enterprise and communication customers in that market or markets is the goal here. Direct customers for the goods or services may not be the consumer or end user, which leads to the need for defining distribution and resale arrangements also.

4.4.1.10 Sales

Sales matches the needs of the customer to the goods or services provided, making the sale by supporting the transaction under which the buyer procures and the seller provides. This would include detailed negotiations on contract terms and conditions, prices, functionality, and delivery schedules. The complete sales process also includes the preparation of invoices, as well as credit control and collection activities, supporting the entire sales process from initial contact with a potential customer to delivery of the goods and services and collection of the monies due.

4.4.1.11 Product and Service Development

Product and service development may take place at two points in time: before launch or market entry and then after launch as the customer base enlarges and the range of products and services grows.

4.4.1.12 Customer Service Management

This is closely related to the operational-management role, but with a specific focus on the management of the interface between the enterprise and the customer. This role would address the management of customer accounts, customer inquiries, and service requests.

4.4.2 Candidate Private-Sector Traveler Information Objectives

There is a wide range of possible objectives for the private-sector organizations and enterprises that can be involved in ATIS. We explore these in this section.

4.4.2.1 Making Profit

This is an obvious choice for the first objective in the private-sector arena and is essential to the long-term success and survival of any enterprise. The difference between what it costs the enterprise to do business and what it brings in by way of fees, compensation, and other payments must be, or become, positive over the long haul. In fact, some schools of thought in business suggest that the difference, known as net income or profit, should be large enough that the return on money invested in the company is higher than the interest you would receive if you put the money in the bank. The logic is simple: If you cannot make more money from the business than you could by putting the money in the bank, then it is not worth the effort to establish and operate the business.

Whether you subscribe to this view or not, profit is a vital source of energy for the private-sector enterprise. Aside from providing your investors and venture capital providers with a fair return, which would encourage them to lend you more when you need it, profit will be the source of funds for future product and service development. These days, profit, financial performance, and stock price performance can also figure in the approach you take to staff retention and motivation.

Of course, it is quite legitimate not to make profit. If your business plan shows a need to invest to develop certain key products or services, or to buy market share and develop a sustainable market for innovative products and services, then you may start out with an initial expectation of short-term loss, leading to longer-term gain. The amount of money you need to inject to keep your enterprise operating in these conditions has sometimes been referred to as the burn rate. This is the amount of external capital you need on a regular basis to keep operating your business, product, and service development plans. This would be analogous to implementing a new transportation system in the public sector. Public funds would be invested initially to implement the system, then the public would accrue benefits over a number of years until the benefits exceed the cost as the implementation passes through the break-even point.

It could be argued that profit is, in fact, the only intrinsic or core value that the private sector seeks in its pursuit of traveler information system objectives. There are a number of instrumental values, however, that are also pursued, and as these directly impact the way in which the private sector acts and interacts with the public sector, they are worthy of discussion.

4.4.2.2 Developing Sustainable Business

This objective is related to the need to make profit and keep making profit, and it also touches on the private enterprise need to identify and manage risks and liability. Risks can be related to changes in technology, changes in the market, or changes in the needs of customers. With regard to technology, it is becoming more and more difficult to be certain that your products and technologies will not be superceded and made unmarketable before you have had a chance to earn a reasonable return on investment. The pace of technology change is accelerating, providing us with a management challenge that forces speedier development and distorts conventional market economics and wisdom.

4.4.2.3 Leveraging the Power of the Internet and Information Technology

Building on the previous objective, most enterprises have a strong desire to leverage Internet and other information and communication technologies that have been developed for other markets and are transferable to traveler information. This quickens the development cycle, reduces research and development costs, and helps to manage the accelerating pace of change. This is, of course, a good lesson for the ITS community to learn. The relatively small size of the markets that make up the world of transportation means that it is very rare to have specific technologies developed especially for these markets. Consequently, more time must be spent in scanning other technology application areas, seeking technologies that can be adapted to and adopted in transportation applications. It is impossible to do so effectively unless we first arm ourselves with a clear understanding of our needs and objectives. Need we say more?

4.4.2.4 Identifying and Managing Risks

Within the context of changing technologies and market conditions, the private-sector enterprise must go beyond technical excellence and encompass sound business-management practices. One of the most important elements of this is the adoption of a sound structured approach to the identification, evaluation, and management of risks. This is not to say that the private-sector enterprise is risk averse, since in our experience most private-sector enterprises will find risk attractive, under the appropriate terms and conditions. The kicker is, of course, in the last part of that sentence. The private sector must devote sufficient resources to risk identification, evaluation, and management to ensure that the terms and conditions are appropriate. One significant effect is the direct relationship between risk and reward. In many cases the private sector will assume risk and liability in the presence of a suitable level and probability of reward and therefore hold the definition of risk and reward as one of the key objectives in participation in a traveler information system deployment or operation.

4.4.2.5 Effective Interfacing with the Public Sector

In many cases, the private-sector participant in traveler information systems will seek an effective and useful interface with the public sector. This is especially true of information service providers in traveler information who are planning to make use of public-sector data-collection and -processing capabilities to supply a data feed to their privately operated information delivery systems. An effective interface would support a clear understanding of what data would be delivered to the private sector and under what terms and conditions it would be supplied. Another dimension of an effective interface would be stability and sustainability over time, underpinned by some form of configuration-control or change-management system on the part of the public sector.

4.4.2.6 Regional and Market-Focused Activities

The market for information services tends to be regional at the smallest end of the scale and national at the other. Traveler information service providers and device manufacturers are seeking a national market for their products and services. Automotive electronic and automobile manufacturers are only able to invest the required level of capital in new product development if there is a national market. Consequently, private-sector participants, with the exception of system integrators and consultants, attempt to secure regional deals for data and information delivery that can be strung together into national coverage patterns. For example, a goal may be to have traveler information delivery capabilities in the largest 25 metropolitan regions in the United States, or every capital city in Europe, or coverage of the entire trans-European road network.

4.4.2.7 Exploiting the Wider Market for Information Services

The private sector cannot ignore the fact that the market for information services is larger than the market for traveler information alone. In fact, the market for traveler information can be considered as a submarket or market segment of the market for information services.

Information providers and manufacturers of information delivery hardware and software have recognized that the market for information seems to work best when a range or portfolio of information services is offered to the consumer. For example, while some travelers may not be prepared to pay for traveler information as a stand-alone service (maybe because the public sector has conditioned us to expect it for free), they may pay for a package or bundle of information services, featuring traveler information as one component. Consequently, there is considerable interest in cobranding and syndication deals to enable the creation and marketing of subscription packages of information, which requires that the information service provider look outside of the world of traveler information and transportation.

4.4.2.8 Identifying and Applying the Best Business Model

The identification and application of the most effective combination of resources, technologies, products, services, business strategies, alliances, market positioning, and partnerships is crucial to the success of a private-sector enterprise. The definition of best may vary from company to company as financial and business objectives differ. Best may be defined as the most profitable, the most sustainable, the most market share, or the most attractive long-term prospects.

4.4.2.9 Obtaining and Preserving a Competitive Advantage

Having invested their own or someone else's money in the development and deployment of traveler information products or services, most private-sector enterprises seek to obtain and preserve some kind of competitive advantage over the competition. This may relate to a time advantage (being in the market first) or a technology advantage (having access to technology that others do not have). When an enterprise has achieved such an advantage, it is very keen to preserve it and is not enthusiastic about collaboration or open information sharing. This type of objective has another facet; that is, avoiding competitive disadvantages and competition from public-sector partners. While they strive to achieve an advantage, they also seek to avoid competitive disadvantage.

4.4.2.10 Developing the Market for New Products and Services

One of the challenges for many private-sector players in the traveler information sphere relates to the effort required to develop the market for innovative products and services. Consumer awareness of the benefits of the products and services and a level of trust and confidence in new approaches and techniques need to be built up over a period of time. This requires a strong approach to communicating the benefits, features, and value of the new products and services through early successes in the market and through the careful development and culturing of key early customers.

4.4.3 Real Objectives Versus Metaobjectives

You may have noticed something missing when you read through the list of public- and private-sector objectives in the preceding sections. There was no reference to establishing and operating a traveler information system as an objective for either the public or private sector. This is because we do not believe that this is a real objective. This may fly in the face of conventional wisdom—our experience tells us that many of the traveler information system deployments around the globe have this as the primary and sometimes only objective. In many cases, the prime justification for investing time, money, and other resources has been the need to establish and operate a traveler information system. In fact, many such systems have been prescribed in terms of specific ways to collect data, process information, and deliver information to the traveler with little or no regard to the identification, confirmation, and agreement of the real or fundamental objectives of the effort. We think this is a serious problem that leads to innovative solutions being ignored and almost certainly handicaps the success of the initiative. We have coined the term *metaobjective* to describe such objectives that seem like real goals, but which are in fact solutions posing as objectives. Figure 4.1 illustrates our point: The metaobjective (establish and operate a traveler information system) is mapped to the true objectives of saving time, lives, and money, reducing stress, and optimizing use of existing capacity. Identification of the true objectives enables full consideration of other options and supports the development of a solution that fully matches all needs, issues, problems, and objectives on both the public and private sides of the fence.

Table 4.2 summarizes the list of private-sector objectives identified and provides a tentative mapping of the objectives to the roles as previously defined.

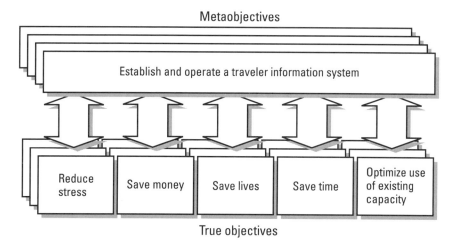

Figure 4.1 Metaobjectives.

Table 4.2
Summary of Private-Sector Objectives

Role	Making profit	Developing sustainable business	Leveraging the power of the Internet and information technology	Identifying and managing risks	Effective interface with the public sector	Regional and market-focused activities	Exploiting the wider market for information services	Identifying and applying the best business model	Obtaining and preserving a competitive advantage	Developing the market for new products and services
Business management	●	●	●	●	●	●	●	●	●	●
Operational management	●		●	●	●	●	●			●
Marketing and distribution	●	●	●		●	●	●	●	●	●
Sales	●				●	●			●	
Product and service development	●		●	●		●	●		●	●
Customer-service management	●	●	●	●		●			●	●

4.5 Summary

In this chapter we have explored the nature of public and private objectives and motivations within the realm of traveler information. We have identified a range of organization types and roles within both public and private sectors and have mapped these roles to candidate sets of objectives. The idea is to give both public- and private-sector readers some insight into the overall drive and direction of each sector, preparing the ground for the identification of common objectives and significant differences on the manner in which to define respective roles and responsibilities within the traveler information supply chain. We will address the identification of common ground and significant differences when we discuss the formation of public-private partnerships for traveler information in Chapter 5. Table 4.3 sets the scene by cross-mapping both public and private objectives. This shows that while there is considerable scope to identify and agree upon common objectives, there are areas of objectives that are not shared. Further work and careful establishment and management of an effective dialog between the two sectors would be required in order to progress to a partnership.

One of the most significant potential areas for collaboration and sharing lies in the identification and quantification of the value of traveler information to the full range of participants through the use of simulation and social decision analysis to guide the formation of effective partnerships. We will address this in detail in Chapter 6 when we explore the possible business models that may be applied to traveler information systems.

Table 4.3
Common Ground Potential Summary

Type	Role	Public												Private									
		Obtaining the best value for public funding	Maximizing the use of legacy systems and sunk investment	Leveraging the efforts of others	Addressing as many organizational objectives as possible	Investing public money in opportunities that are appropriate for common funding	Leveraging the motivation and desire of the private sector	Traveler behavior change	Jurisdiction-, mode-, or corridor-focused activities	Promotion of a particular travel mode	Reduce intermodal travel times	Improving quality of service	Maintaining an appropriate level of control and management direction	Making profit	Developing sustainable business	Leveraging the power of the Internet and information technology	Identifying and managing risks	Effective interface with the public sector	Regional and market-focused activities	Exploiting the wider market for information services	Identifying and applying the best business model	Obtaining and preserving a competitive advantage	Developing the market for new products and services
Public	Operations professionals	•	•	•					•		•	•		•	•	•		•					
	Transportation-engineering and -planning professionals	•	•	•	•	•	•	•		•	•			•	•	•		•					
	Procurement specialists/contracts professionals	•	•	•	•	•						•	•	•	•	•	•		•	•			•
	Public-information, marketing, customer-service professionals	•	•	•	•	•	•	•	•	•	•	•		•	•	•	•	•		•			•
	Political professionals	•	•	•	•	•	•	•			•	•		•	•	•	•	•		•			•
Private	Business management	•		•	•	•							•	•	•	•	•	•	•	•		•	•
	Operational management	•	•		•	•	•						•	•	•	•	•	•	•		•		•
	Marketing and distribution	•		•	•	•							•	•	•	•		•	•	•	•		•
	Sales	•		•	•	•							•	•	•			•	•		•		•
	Product and service development	•		•	•	•							•	•		•	•		•		•		•
	Customer-service management	•		•	•	•							•	•	•	•	•		•			•	•

References

[1]   "Transportation Equity Act for the 21st Century," http://wwwcf.fhwa.dot.gov/tea21/ summary.htm.

[2]   Department of Transport and Regional Services, Australia, "Mission Statement," http://www.dotrs.gov.au/land/General/ourmission.htm.

[3]   Rescher, N., *Introduction to Value Theory*, Lanham, MD: University Press of America, 1982.

[4]   "ATIS U.S. Business Models Review," prepared for U.S. Department of Transportation ITS Joint Program Office, HOIT-1, Washington, DC 20590, by Rick Schuman and Eli Sherer, PBS&J, November 15, 2001.

5

The ATIS Supply Chain

5.1 Learning Objectives

After you have read this chapter you should be able to do the following:

- Explain, in detail, each step in the advanced traveler information supply chain;
- Understand the practical significance of the advanced traveler information supply chain;
- Define the wider domain within which ATIS are deployed and operated;
- Appreciate the value of supply-chain management for traveler information applications.

5.2 Introduction

When we introduced our operational concept for ATIS in Chapter 2, we used it as a means of illustrating how a range of technologies can be applied to ATIS from the perspective of technical integration. The concept shows how different subsystems, comprised of information technologies, linked together by telecommunication technologies can function as a single system for data collection, information processing, and information delivery. We also touched on the fact that the configuration of the operational concept incorporated several assumptions regarding operational and management responsibilities for subsystems and the overall system. We assume that the public sector will collect and share the required data and that the private sector will have a significant role in the delivery of the information. It is obvious that the identification and consideration of operational or business options for ATIS cannot be separated from the technical aspects and that there is a wide range of options and permutations to take into consideration. It would be of considerable value to have a simple, structured view, or reference model for ATIS as a tool for identifying, assessing, and selecting technical and business options. This would help to ensure that the identification and evaluation of business and technical options is thorough, complete, and carried out in a systematic manner.

In this chapter we attempt to provide such a tool in the form of a basic coherent structure to assist with the task of understanding ATIS and managing the overall business process. We call this the advanced traveler information supply chain as it defines the major activities required to convert resources (time, money, data, systems, and people) into the desired value for the traveling public in the form of traveler information products and services. We based our model on the type of approaches that are taken in logistics and process planning [1, 2], drawing from experiences in these fields and in the field of business process reengineering. Combining this with an analysis of the major advanced traveler information deployments around the world and the framework provided by the U.S. National Architecture for ITS [3], we have distilled the advanced traveler information process into a simple six-step supply chain that can be used as a checklist and reference model. We have also drawn inspiration from some pioneering work being carried out in Europe [4] with regard to the definition and understanding of the information supply chain.

The term *supply chain* may be a bit strange and new in the world of the ITS specialist and transportation professional. It is typically used in logistics and manufacturing to describe the process by which raw materials are converted into manufactured goods, distributed, marketed, and sold to customers. All we are doing here is using a term and technique from another discipline area and applying it to advanced traveler information. We like the term since it conveys the message that traveler information is about supplying something to the traveler and that a chain of events or activities must be coordinated and synchronized if it is to be carried out effectively. It would also be appropriate to refer to the chain as the advanced traveler information value chain as it enables us to define the various points along the process where value is added or removed. This is a common way of describing the business process when conducting reengineering activities.

One of the first steps in improving the management of just about anything is to obtain a clear picture of the process to be managed. It is critical to know how the process works and to identify its strengths and weaknesses. This is a prerequisite to the identification and implementation of performance monitoring and measurement. Only then have you equipped yourself with the tools and information required to effectively manage the process. In the course of defining and describing the process, it is also possible to relate the steps being supported to objectives, tracing the effectiveness of the process in achieving the desired results.

As many ATIS involve a variety of resources and activities and could potentially involve a wide range of different organizations, we believe that the adoption of a systematic approach to process definition is absolutely essential. We believe that the application of business process definition and management techniques to ATIS initiatives holds the key to increased effectiveness and improved efficiency.

This chapter introduces the concept of the advanced traveler information supply chain and explores the various activities that comprise the chain. After describing and defining the chain, we then make use of it as a tool in Chapter 6 to explore the forms of public-private partnerships that may be possible in the realm of traveler information systems, as well as the wider context within which the process takes place, and to define issues and challenges.

5.3 The ATIS Supply Chain

Figure 5.1 illustrates the traveler information supply chain as we have defined it, showing the six major activities to be supported and the relationships between them.

Let us take a look at these major activities one by one. Note that we have numbered these in increments of 100 as we expect additional steps to be added as further experience is gained in the application of the chain. This also makes it easier for you to customize the basic model for your needs, while retaining the original numbering for reference purposes.

5.3.1 Step 100: Building the Data Infrastructure

The first activity in the traveler information supply chain revolves around the creation and maintenance of an infrastructure capable of providing the data required by the traveler information system. When we talk about a data

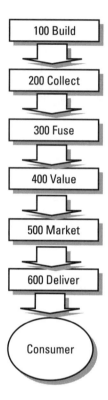

Figure 5.1 The ATIS supply chain.

infrastructure, we envision a combination of automated and manual resources composed of automated sensors, communications networks, and manual anecdotal data collection. ATIS will tend to have a heavier emphasis on the use of automated, objective data collection, but may make some use of anecdotal subjective data to fill gaps in automated coverage or provide additional background data. Anecdotal data could come from a variety of sources, such as spotter planes, highway patrol officers, the public at large, or fire and other emergency-management services. The data received from these types of sources tends to consist of unstructured, narrative accounts of traffic or transportation-network conditions, with subjective assessments of levels of severity and anticipated durations. A typical data infrastructure for a traveler information system could include invasive and noninvasive sensor technologies, a communications network to carry information from the transportation network to a central processing facility. This facility is likely to be staffed by at least one operator capable of receiving anecdotal and human-sourced data to be parsed into the database alongside the automated data. A sophisticated data input tool that supports the input of the anecdotal data by the operator usually supports this manual stream. This entire infrastructure needs to be planned, developed, designed, and implemented. The establishment of procedures and agreements to source the data would also be part of the infrastructure-building activity.

Ideally, a thorough understanding of the needs, issues, problems, and objectives that we discussed in Chapter 4 will directly drive these activities. This should lead to a definition of the data required to enable the traveler information system to provide and support the services that will in turn attain the desired objectives. Based on this understanding of the data requirements in terms of types of data, accuracy, and completeness of coverage needed, it is possible to design and develop a data infrastructure that effectively and efficiently provides the required data. Technologies are evaluated and selected from the range discussed in Chapter 3 and supplemented with the human-centric anecdotal data discussed in this chapter.

In addition to the selection and design of the technology elements, this part of the traveler information supply chain or process would also include the development and implementation of appropriate operational rules and procedures for operating the data-collection system. Although these rules will be applied as part of the next step in the value chain, we need to consider them now since the design of the data-collection facilities will be influenced by how we plan to use them.

Another important consideration in the development and deployment of the traveler information systems data-collection elements is the integration with the data-collection needs of other ITS systems. Obviously, it would not make much economic or engineering sense to deploy parallel duplicate data-collection capabilities for traveler information, traffic management, and other transportation management and planning needs. Therefore, the data-collection infrastructure should be designed, developed, and deployed within a detailed understanding of bigger picture data needs. These needs relate to the data requirements of other ITS and transportation applications that form integrated approaches to transportation-network management and operations. In the United States in particular, there is growing recognition of the value of a so-called "national infostructure" comprised of sensors and communications facilities that provide the basic data required for effective transportation-network management and that feed the traveler information services needed to promote effective end user use of the facilities. We will discuss this in detail in Chapter 7 when we address the position of traveler information within the regional ITS future big picture.

The establishment of the data infrastructure represents a major investment decision on the part of either the public or the private sector. The choice of data-collection technology—whether manual anecdotal, automated sensor, or a hybrid of the two—is a fundamental business decision to be made. If sensors are to be deployed, the exact content, format, and structure of the required data and the nature of the information to be derived from such data must be determined. The granularity (i.e., spacing or density) of the sensor network must be related to the required level of detail to support the information stream to be subsequently delivered.

An effective and successful approach to the planning, development, design, and implementation of a data infrastructure for transportation-network data collection must be driven by objectives. There needs to be a clear idea of the types of data required in order for the system to deliver the user services desired. This requires a thorough requirements and systems analysis to be performed, which would lead to the identification and selection of suitable sensing technologies, manual data-collection methods, and communications facilities. One should begin with two fundamental questions:

1. What data will our system need to deliver the services we want? Vehicle speeds, flow and volumes, location and classification of incidents, journey time estimates, travel-demand patterns (origin, destination), traffic-signal times, network status, time to next transit vehicle, timetable and schedule data, cost of travel and fares, boarding and alighting locations, intermodal interchange possibilities.

2. How can we collect the data required? Other existing systems, hard-wired or invasive sensors, noninvasive sensors, vehicle sensor devices, network sensor devices, and automated or manual data-collection techniques.

When pondering these questions, we should also think ahead and consider the data-collection procedures that will be adopted when actually operating the data-collection infrastructure. As previously discussed, the data will be collected in both manual or human-centric ways as well as by fully automated data-collection systems composed of sensors, communications, data processing, storage, and retrieval systems. Consequently, the procedures must be defined for the technology, or technical aspects of the system, and for the people, or organizational aspects. From the technical perspective, the system architecture to be developed should define all of the major subsystems and the communication links that will help them to work as a single coherent system. From an organizational perspective, a set of joint operating rules that define roles, responsibilities, and management arrangements needs to be developed and agreed upon. We often identify a third major perspective: that of the commercial or enterprise arrangements. This defines and agrees upon rules or business models that describe how the revenue, profit, subsidy, and risk will be managed.

5.3.2   Step 200: Collecting the Data

The second step in the traveler information supply chain involves the operation and management of the data infrastructure defined in the first step. This involves a range of activities, including the following.

5.3.2.1   Technical Processing Arrangements

These involve the definition of the operational concepts and procedures that will be applied in the operation and management of the data infrastructure. The data-collection facilities need to be operated. These include all of the communications and information-processing hardware, the operating systems, and the software. Any hybrid (part automated, part human) data-collection procedures need be executed. This includes the recruitment and training of operations staff and the ongoing management of the hardware, software, data, and people.

If the system is new, then this activity might also include acceptance testing and commissioning as the new system is handed over from design and development to operations. Later, the work will also involve managing the communications and sensor network, maintaining the data-collection infrastructure, and overall maintenance management and planned replacement of system components. There may well be a large number of interfaces to be supported and managed, between the data-collection system, other data sources, and external partners.

5.3.2.2   Operational Arrangements

A thorough analysis of roles and responsibilities should be used as the basis for determining the operating rules for such interfaces. This may require the development and agreement of formal memoranda of understanding or other interagency and interorganizational agreements. If the other partners involved in the data-collection process are from the private sector, then it may be necessary to add other items to the agreement that define how liability is apportioned and describe how revenue will flow between partners.

The collection of data to provide travel information fees may not be mission critical or a central focus in the operation of a public-sector transportation agency. It may, however, be an important thread in the fabric of an effective transportation-management approach. Travel data can be viewed as part of the portfolio of mission-critical data that the effective public transportation agency must collect and process. Consequently, when addressing the data-collection needs of the traveler information system under consideration, the smart agency will also address the wider needs of data collection and information processing for the whole spectrum of transportation planning, development, and operations. A focus on travel data and information should lead to at least a brief review of all data-collection and -processing activities, which include the following:

- Traffic-management data;
- Transit-management data;
- Bridge-management data;
- Pavement- and maintenance-management data;
- Transportation-planning data.

There is substantial potential for coordinated data collection by planning and executing data collection and use activities across departments and applications. Data could be considered as a shared resource that supports the business processes of each separate department. This would, of course, require the integration to some degree of various data-collection activities and the convergence of spatial referencing systems for the data.

5.3.3   Step 300: Fusing the Data

Data fusion involves the collation and integration of data from a number of different sources. The data to be collected by utilizing the data infrastructure, under the operation of the data-collection system, is usually collected and stored by some form of database-management system. A data warehouse supporting sophisticated data processing, storage, and retrieval capabilities would most likely be deployed. This part of the chain involves a set of activities aimed at taking data from multiple sources and collecting it into a single coherent data source. This requires that the data be verified, cleaned, corrected, made complete, and formatted for future management and retrieval. The data-fusion process also encompasses the conversion of anecdotal, verbal, and subjective data into objective, structured, textual, or numerical data. The work carried out on the data will also include the conversion from multiple-location or spatial referencing systems to a common single-location referencing standard.

In addition to collecting and collating the data into a single manageable source, fusion may also involve the addition of data from external sources, or the use of independent data to verify the data coming from the data infrastructure. Bringing all the data to a single unified spatial referencing convention may also be part of the fusion step, although it may also be possible to leave the data in different referencing standards and develop correspondence tables or relationships between them to support a unified data retrieval process.

5.3.4 Step 400: Adding Value

In this part of the chain, the unified data source is converted to information and packaged or combined with other information to provide value propositions that can be (1) marketed to consumers for a fee, (2) provided for free, or (3) provided at no charge to the consumer with advertising or sponsorship underwriting or support for the cost of delivery. Traveler information may be combined with weather, sports scores, or stock quote information to provide subscription packages for direct retail sale to consumers, or for wholesale distribution to downstream information services businesses and partners. For example, traveler information may be supplied to a smart yellow pages service provider and form part of the content for an interactive voice response system offering multiple information streams, or to a Web site acting as a multipurpose portal for entry to the Internet.

These activities can also include the customization of information to user needs and desires. Information can be converted to different textual and graphical formats, filtered, transformed, and modified to suit consumer needs.

It is at this step in the process that information products and services are developed and delivered to the information consumer. The steps above this one in the supply chain supply the raw materials from which these products are made. This would be akin to lumber entering the carpentry shop and being converted into furniture, for example. Continuing the analogy, the smart carpenter will have made sure that he or she has a good understanding of the needs and wants of the customers who are buying the furniture. There would not be much point in making patio furniture in the dead of winter, for instance. Therefore, the information product or service developer and deliverer will have a fairly detailed understanding of who the customers are, what they want, and when they want it. Many of the players in this step of the process will have clear commercial objectives that bear little or no relation to the attainment of public transportation policy objectives. While there may be some shared objectives, the public sector may have little or no control over many of the players in this part of the process.

Consider the example of the information service provider who has a supply contract to provide traveler information content to the in-vehicle information systems being installed on the production line by a major automotive manufacturer. The main thrust of the information package to be supplied will undoubtedly focus on providing the driver with a personal optimum set of information that will give the driver a competitive advantage and hence provide a value proposition, making the sale of the unit and the service viable.

Now consider the objectives of a local government transportation agency within whose jurisdiction the in-vehicle information system will operate. It is likely that this public-sector agency has a different objective: that of identifying and supporting community optima, rather than personal ones. In such a case, the final delivery of the traveler information is only of benefit to the public sector if personal and community optima coincide. Structuring roles and responsibilities within the overall framework of the traveler information supply chain is an important part of the development of public-public and public-private partnerships designed to create win-win situations in which transportation policy objectives are achieved within a commercially viable business framework. We will discuss various ways to do this in Chapter 6.

5.3.5 Step 500: Marketing the Information

Irrespective of whether the traveler information is provided free of charge by a public agency or on a commercial basis by the private sector, marketing activities are an essential part of the process. The target users for the information must be made aware of some basic information regarding the use of and access to the information as follows:

- How to access the information;

- How to use the information;

- Benefits and value of making use of the information;
- Cost of accessing and making use of the information.

The marketing activities should also include steps to understand how the consumer uses the information and what the real and perceived value of the information is. This should lead to the identification and development of new information products for the market. Advertising and promotion of the traveler information services is likely to play a pivotal role in the development of the customer base for the services. Like any product or service offering in the market, the benefits and features of the traveler information need to be explained, defined, and described to the target groups of users. A value proposition for the user needs to be defined and communicated if the users are to make a positive decision to make use of the services. When the information stream has a price associated with it (in terms of a user fee or subscription), then the cost to the user needs to be justified by illustrating how the benefits of using the service outweigh the costs of doing so. Even in cases where the information is to be delivered free of charge as part of a public service offering, it is vital to carry out this cost versus benefits analysis and exposition. In such cases, the cost of making use of the services may be characterized as time investments required to access the information and make use of it. The marketing and advertising industries define a process that the potential user of the service or product must be supported through in order to get to the sale, or the use of the service. It is referred to as AIDA since the four steps are awareness, interest, desire, and action.

- *Awareness:* The potential user of the product or service must be aware that it exists and broadly familiar with the features and capabilities available.

- *Interest:* Once the user is aware, it is necessary to generate interest in finding our more about the product or service and instill a curiosity about what it is and what it can do.

- *Desire:* As the user's familiarity with the product or service grows, the information provided must support the development of a desire to acquire or make use of the product or service. As stated earlier, a solid, clear-cut value proposition must be developed and effectively communicated to the target users.

- *Action:* The final part of the sales cycle is when the user decides to take action and acquire the product or service. It is important at this point in the process to make it as easy and simple as possible for the user to complete the cycle and access the product or service.

Marketing or promoting the traveler information products and services effectively also requires a detailed and current understanding of user needs. It is vital that the offerings in the market closely match the needs of the intended user, making it feasible to support the AIDA cycle and develop a solid value proposition. This means that the marketing and promotion of the traveler information should also include evaluation and assessment of user needs. As these change, the information on the new needs can be fed back into the adding value step in the traveler information supply chain, where new product and service combinations can be defined to meet the new needs. Marketing can also encompass the measurement and assessment of the real benefits that users are achieving through the use of the traveler information. This generates information that can be utilized to support the value proposition but also provides another feedback possibility for the earlier steps in the traveler information supply chain. Real-world benefits and impacts on the travel behavior of the users are a vital part of the justification for public-sector involvement in the traveler information supply chain. It is very important that information on the actual benefits and impacts of the traveler information be compared with the intended benefits and impacts that drove the development of the system in the first place. This information may lead to the redefinition of requirements, the reassessment of data infrastructure and collection needs, and a review of the respective public and private roles in the traveler information supply chain.

5.3.6   Step 600: Delivering the Information

Delivering the information is also referred to as distribution, and this occurs at the downstream end of the traveler information delivery or value chain. The information is pushed through multiple information delivery channels for delivery to the consumer of the information. These channels may be operated by the public or by the private sector. For example, the public sector may channel traveler information through dynamic message signs, interactive voice response systems, or information Web sites to the traveling public. Publicly operated information delivery channels are typically free of charge, with funding derived from common funding sources, such as local, state, or national tax revenues. Private-sector information delivery channels may include both static and mobile Internet channels, TV, radio, pagers, and cellular telephones. There is a growing market for in-vehicle information devices and information services that would also be part of the private-sector information delivery mechanism.

Looking at information delivery within the context of a regional traveler information system, the range of potential delivery channels needs to be defined and understood. There will be a number of publicly operated information delivery channels related to the transportation process. These may include transportation agency Web sites or roadside dynamic message signs. There may also be a large number of privately operated information delivery channels aimed at the delivery of a wide range of information services, with transportation and traveler information as a subcomponent, or part of a larger information portfolio. Examples of these would include cable TV and private information service providers.

Effective use of the many delivery channels requires the development of a business approach that assesses and evaluates the different delivery possibilities in a structured, coherent manner. Such an approach would include the functions described in Sections 5.3.6.1 to 5.3.6.3.

5.3.6.1 Delivery Channel Identification and Characterization

This involves the identification of the full range of delivery channels available within a region and an assessment of the features and characteristics of the channels. This would include an assessment of the cost of using the channel, the number of people with potential access to the channel, and the suitability of the channel for delivering the information to the right people, in the right format, at the right price, at the right time, and in the right place.

5.3.6.2 Selection of Appropriate Channels Based on Support of Objectives and Willingness To Pay

Building upon the understanding of what the channels can do, a selection would be made by matching objectives and needs to channel capabilities and characteristics.

5.3.6.3 Definition of Shared Resources Opportunities

In order to take full account of the possibilities for cost sharing and synergy between the public and the private sectors, the business approach should also include a full examination and assessment of the possibilities for shared resource approaches.

5.4   Applying the Advanced Traveler Information Supply Chain to Our Operational Concept

In order to illustrate the application of the advanced traveler information supply-chain tool, we use it to structure our operational concept as described in Chapter 3 and at the beginning of this chapter. Figure 5.2 shows a mapping of the operational concept using the chain.

Figure 5.2 uses the vertical axis to indicate what step in the ATIS chain is being supported and the horizontal axis to indicate who is supporting it. We often refer to this as a what-who diagram. Note that we have combined the steps in the advanced traveler information supply chain into pairs to make the diagram clearer since there was no need to differentiate these steps in our operational concept. The diagram depicts our operational concept as described in Chapter 3, illustrating the technical linkages but also highlighting the business rules of business model assumptions that are implicit in our choice of operational concept. This is significant as we are now able to explore and examine the relative roles and responsibilities of different organizations and entities rather than just the technical requirements. Our operational concept features public-sector dominance in the early part of the chain (Steps 100, 200, and 300) and private-sector dominance in the later part of the chain (Steps 400, 500, and 600). This assumes that the public agencies involved—traffic, transit, and emergency management—will build and operate the data-collection infrastructure required to provide the data to feed both advanced traveler information and traffic-management applications and share this information with the private-sector information service provider through some predetermined partnership or contractual relationship. Data fusion is also supported by the public sector through the establishment and operation of a central data warehouse that collates and fuses data from multiple sources for reuse in the public context and relay to the information service provider. Apart from roadside dynamic message signs, traffic signals, and transit information kiosks (which are all considered to be part of the transportation-management system as are traveler information delivery mechanisms), the delivery of traveler information is handled by the private sector through the establishment, management, and operation of Web sites and other information delivery channels, such as TV and radio broadcasts.

Note that the use of the six-step advanced traveler information supply-chain reference model forces a systematic review of the business and operational assumptions that have up to this point been implicit. This review may also uncover alternative approaches and lead to a revision of business approaches and operations. For example, if the public agencies have little

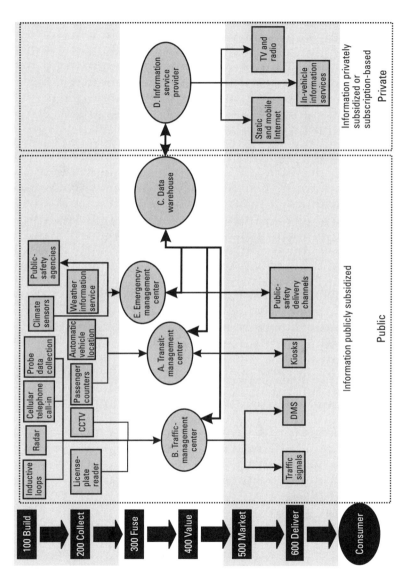

Figure 5.2 ATIS operational concept mapped to roles and responsibilities. (*After:* [5], which provides an excellent overview of the business models that have been applied to traveler information systems in the United States.)

data infrastructure in place and require time and money to establish one, they may decide to make use of private-sector data-collection resources, such as spotter planes and field observers as an interim measure. This leads us to another aspect of the supply chain, the temporal dimension. It is entirely probable that sector dominance in the different steps in the chain will be dynamic, with an initial configuration or business model being transitioned to another one as relative public- and private-sector capabilities change over time. This structured approach to the identification and agreement of respective roles for each sector is of great importance in achieving efficiency and effectiveness. The use of this simple supply-chain model provides a good way to approach the analysis and communicate the resulting plans with key decision makers and potential partners. As we will explore in Chapter 6, the supply can be used as part of the initial planning for ATIS to outline the proposed business model or approach, to identify and define respective roles and responsibilities, and to form the basis for meaningful dialog and communications leading to the development and agreement of detailed business models and specific partnering and procurement approaches.

5.5 Traveler Information Supply-Chain Management

As we stated in Chapter 1, our primary interest in this book is to address the issues and challenges associated with the open traveler information system. If the system is limited enough in scale that the entire chain can be accommodated and supported within a single organization, then this would be defined as a closed traveler information system. An example of a closed system would be a passenger information system for a single public transit agency. Such a system would make use of data collected only within the organization, and the information would be delivered only to users of the transit system; all processing, adding value, marketing, and delivery would be carried out internally as well. This would also be the case if a private-sector organization decided to support all activities in the supply chain as part of an exclusively private operation, supported entirely within the confines of the corporation. While the traveler information supply chain applies to such closed systems and is useful in guiding the planning and development of such systems, our real interest lies in providing a frame of reference for the larger-scale, open systems. These tend to be regional in coverage and undoubtedly involve collaboration between agencies and organizations. You probably understand this already, but we will make the point anyway since it is so important. In terms of an open traveler information system, the traveler information supply chain consists of a number of steps or activities that bridge organizational boundaries and span different domains. Unlike conventional transportation projects deploying asphalt, concrete, and steel under the auspices of projects and procurements that are entirely within the jurisdiction and control of public transportation agencies, traveler information systems spill out into a wider world. Public agencies find themselves in a new paradigm where advocacy, influence, and effective communication and partnering are prerequisites for success and effectiveness. It is almost impossible to address the development and operation of an effective and successful traveler information system without full consideration of the management needs of each step in the entire process. The best results are to be had when this process is considered in its entirety and managed as a cohesive single system. Figure 5.3 illustrates the advanced traveler information supply chain within the wider context of other domains.

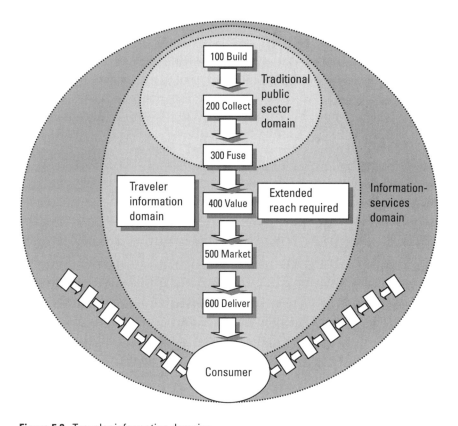

Figure 5.3 Traveler information domains.

Let's take a quick look at some examples of the types of players we might encounter in each zone of the diagram. In the inner circle, we would, of course, expect to encounter public transportation agencies including federal, state, and local highway authorities and transit agencies. We may also bump into one or more traveler information service providers collecting and processing their own data in parallel with the public transportation agencies.

In the middle zone we would expect to see the public transportation agencies extending their reach to include the marketing and delivery of traveler information and the traveler information service providers. We would also expect to see the automotive manufacturers, automotive electronic manufacturers, and software providers responsible for the development and sale of in-vehicle information systems and services.

In the outer zone these players would be joined by many information product and service providers, some dedicated to transportation and some not. These would include so called p-commerce (position commerce) players, who support financial transactions relating to the current location of the buyer, information service providers for nontransportation information, such as stock quotes and sports scores, and device and appliance developers and vendors, who offer a wide range of information delivery devices, such as personal portable computers and enhanced wireless telephones.

Going beyond the management of projects and programs within our jurisdiction and control requires a supply chain-management approach, such as those adopted by the logistics industry. In addition to the requirements of each step in the process as described above, there are a number of overarching process-management considerations that can only be successfully addressed through the adoption and use of a holistic, total process-management approach. In business engineering terms, we are defining our business process and identifying ways to manage it and apply information and communication technologies in the most effective way.

Let us briefly identify the process-management issues and discuss how to address them.

One of the primary overarching issues or process-management concerns is the definition and application of quality and effectiveness management. Starting at the beginning of the chain with the identification of the data needed to support the upstream information-processing and delivery systems, we can identify, describe, and define a series of performance measures to be applied to the management and monitoring of the operation of each step in the process. Examples of performance measures include the following:

- How old is the information?
- How often is it updated?
- What is the reliability or confidence level of the information?
- What is the value to the consumer?
- Did the consumer get what was asked for?

It is especially important to define, agree, and document performance measures (measures of effectiveness) for each step in the process, defining the quality of the data or information expected as input to and output from each step. Since there may be many different organizations involved in the supply chain, with both public- and private-sector roles and responsibilities, it is particularly useful and valuable to have this level of quality definition in support of effective partnership and interfacing between departments, agencies, and private-sector organizations. It may even be possible to make the definition of performance measures a central part of a procurement contract between parties in the supply chain. Here again, we can refer back to the original public-sector transportation policy goals as one of the primary sets of performance measures. If goals, such as reduced travel times, congestion avoidance, and smoother, easier modal and route interchange are being attained, then it is also likely that the value proposition being developed by private-sector players is strong enough to attract subscription and advertising-supported users.

Another issue relates to the definition of appropriate roles and responsibilities for each player in the supply chain, recognizing that the choice of player will significantly impact performance and effectiveness and that steps in the chain have consequential impacts on each other.

One way to consider the cooperative operation and management of the whole traveler information supply chain is the concept of the virtual corporation. This is a collection of resources from multiple partners and organizations, aimed at the development and support of a single business process in the most effective fashion. The various resources are blended, structured, and managed as a single enterprise supported by the appropriate agreements and contracts. Resources may come from several different companies and public agencies with little in common other than the pursuit of benefits and financial returns directly related to the defined business process or supply chain. In order to assemble and operate such a virtual corporation, it is necessary to define the business process as we have defined the traveler information supply chain at the beginning of this chapter. Building on this as a framework, it is necessary to define and agree upon the shared objectives and mutually exclusive goals of each partner participating in the supply chain. Then the supply chain as it currently exists can be reviewed, understood, and a desired future state can be defined. Arrangements, actions, and resources required to migrate the supply chain from the existing state to the desired future state can then be identified, planned, and agreed upon. This usually takes the form of a traveler information system business plan, or a business process improvement or reengineering program. The ingredients of the desired future state would take full account of the needs, issues, problems, and objectives of all the major players in the supply chain, recognize the capabilities of each player, and be based on best-practice information gleaned from a benchmarking exercise. This would compare the current business process or supply chain with the best examples from other cities, states, or countries. The performance measures we defined to be utilized to monitor the performance of each step in the chain would be given target values based on this benchmarking exercise. The desired future state would also be designed to optimize the value that flows from one step in the chain to another and from the beginning of the chain to the end. In the course of the operation and management of the supply chain, we would expect value to be manipulated in three ways, each of which requiring proper and positive definition and management.

1. *Value creation:* This is the creation and addition of value to the chain through the application of resources; for example, the processing of data into meaningful information by deriving summary measures and tabulating or graphing the output. Taking raw data and manufacturing customized travel reports for specific users would fit into this category also.

2. *Value refinement:* This involves adding to the value of the proposition or the offering through a detailed understanding of the needs of the users. This may or may not involve the application of significant additional resources, but also may be simply a delivery of the right information at the right time in order to maximize the value of the offering.

3. *Value migration:* This occurs when value developed or created in one step in the supply chain is realized in another step. This could mean that one player in the supply chain has expended resources to create value or has been able to refine value, while another player in the chain has been able to capitalize on this by realizing

that value. For example, a public agency may have developed, operated, and managed the data-infrastructure, data-collection, and data-fusion steps in the traveler information supply chain, while a partner further down the chain has been able to unlock the value created in these previous steps by actually delivering the information to the user. When value migration is identified in a supply chain, it is likely that there is a strong need for a client-vendor contractual relationship or a public-private partnership agreement to define and agree upon the value being transferred and define suitable compensation for the value creator to provide to the value receiver.

Having defined at least the basic traveler information supply chain, we have empowered ourselves. We now have the ability to review each step or activity in the supply chain and consider the following questions:

- How do we support the process most effectively?
- What activities are best carried out by the public sector?
- What activities are best carried out by the private sector?
- How do we measure the effectiveness, efficiency, and quality of our performance?
- What resources do we need?
- How best should the resources be applied?

The allocation of resources, roles, and responsibilities to the traveler information systems supply chain results in the definition of what we call a business model. In the next chapter, we will take a close look at a number of different business models that either have been used or are feasible for use in the traveler information context.

References

[1]   Kuglin, F. A., *Customer-Centered Supply Chain Management: A Link-by-Link Guide*, La Verque, TN: Ingram Publishers, 1998.

[2]   Coulson-Thomas, C., *Business Process Re-Engineering: Myth and Reality*, London, U.K.: Kogan Page, 1997.

[3]   National Architecture for ITS, U.S. Department of Transportation, http:// www.itsa.org.

[4]   Chen, K., and J. C. Miles, (eds.), *ITS Handbook 2000: Recommendations from the World Road Association (Piarc)*, Norwood, MA: Artech House, 1999.

[5]   Hallenbeck, M., "Choosing the Route to Traveler Information Systems Deployment," U.S. Department of Transportation, ITS Joint Program Office, Washington State Transportation Center (TRAC), 1998.

6

ATIS Business Models

6.1 Learning Objectives

After you have read this chapter you should be able to do the following:

- Define the term *business model*;
- List the range of basic traveler information system business models;
- List a range of enhanced business models for advanced traveler information systems;
- Relate enhanced business models to a selection of specific implementations.

6.2 Introduction

One of the crucial aspects of a successful approach to traveler information systems planning, development, and deployment lies in the definition and execution of a coherent, objectives-driven set of actions involving the deployment and investment of both public- and private-sector resources. There are many possible permutations and combinations of actions and investments that can all lead to the deployment and operation of a successful traveler information system in the right circumstances. The art of the matter lies in understanding the full range of options and selecting the right one for the particular circumstances. There is no single correct answer. The definition and understanding of the traveler information system supply chain (see

Chapter 5) provides us with a framework within which we can explore, discuss, and evaluate the relative roles and responsibilities of the public and private sectors. We can use this information to define and explore numerous combinations or roles and responsibilities, examine the value created and transferred at each step, and determine the most suitable business model for developing, implementing, and operating the proposed traveler information system. This chapter provides an overview of the range of business models, or combinations of roles and responsibilities that are feasible within the traveler information context and some that have been adopted in practice. We will also discuss these business models in relation to lessons learned and practical experience gained in the application of the models in various locations internationally, as well as the advantages, disadvantages, and relevant characteristics of each model. Since the development and adoption of a business model is likely to be different for the public and private sectors, we take a look at this and at the relative advantages and disadvantages from both perspectives.

Although our objectives in developing the materials in this chapter are not to try to tell the public and private sectors how to run their businesses, we have observed a lack of efficiency in the structure of the business models for many prior implementations of ATIS. Results have been impressive but may not have taken full advantage of the resources and capabilities of both public and private sectors. Examples of duplication of resources and overlapping investments are apparent in many deployments, particularly with respect to the development and operation of the data-collection infrastructure and the establishment and management of information delivery mechanisms. For example, there are instances where both public and private sectors operate competing information delivery channels in the same region, while both share the same data sources. There are other situations where a high degree of overlap exists between public and private data-collection facilities. This presented us with a challenge when defining and describing possible business models for ATIS. On the one hand, we want to make sure that the models we develop and describe reflect the needs-and-objectives-driven approach we have advocated throughout the book. On the other hand, we want to provide business models and approaches that reflect the way in which current deployments have been structured, planned, and operated. To address this we have defined two sets of business models in this chapter. The first set, which we have defined as basic models, features the user-needs-driven approach and provides solutions that are optimized in terms of simplicity and clarity with respect to relative public- and private-sector roles and responsibilities. The second set, which we have defined as enhanced business models, have more detail and reflect the current state of practice. This approach also supports the logical progression that we have created in the book, moving from a description of advanced traveler information systems from a technology and technical perspective to a detailed view of the business and commercial aspects. We encourage you to study both sets of models in order to develop an understanding of the needs-driven approach and also to understand the nature of the business models associated with many major deployments. We want to ensure that both public- and private-sector participants in ATIS have as full a set of information as possible on the options that are available, their nature and characteristics, and how others have approached the subject. We also want to provide information to the public sector on how the private sector may think about ATIS and vice versa.

Through an exploration of the theoretical possibilities and the provision of practical business models examples, this chapter provides an overview of the issues and challenges associated with traveler information systems business operations. We hope that this will provide you with an overall frame of reference within which you can develop your particular approach to the subject. We also intend that this chapter be used as a reference source for exploring possible public- and private-sector business-planning approaches and business models. This should support the selection and application of the most appropriate approach for the respective sectors for specific regional and national contexts, maximizing the effective application of resources.

6.3 What Is a Business Model?

Since we are going to make use of the business models concept, we thought it would be a good idea to explain what our understanding of the concept actually is. The term *business model* is heavily utilized in the private-sector world of business and commerce in relation to the overall direction and motivation of an enterprise. The business model for an enterprise explains to potential investors how the organization will turn resources into value for investors, owners, and customers. It explains how the enterprise will generate revenue, make profit, and sustain business by virtue of its positioning in the overall value chain or supply chain. The business model for an enterprise defines the way it will do business, how it will interact with other participants in the supply chain, and what capabilities or features will make it unique or provide it with a competitive advantage. It defines how the enterprise plans to develop and sustain profitable business through positioning in the supply chain. To illustrate the concept, the following are some elements that might be found in a typical business model.

6.3.1   Identification and Definition of Proposed Position in the Supply Chain

This involves the definition and description of the place in the chain that is to be occupied by each participant, the proposed target market, and customers. Having identified the step or steps in the advanced traveler information supply chain that are to be addressed and supported, a detailed view is developed of the inputs required, the specific steps or activities that will be supported by the actions of that participant, the proposed outputs, and the target market for those outputs. There are two ways to approach this: market-driven and technology-driven. Taking a market-driven approach to this would entail starting with a thorough review of the proposed target market in terms of the needs and characteristics of the consumers in the market. This leads to a definition of the outputs (products and services) that are required and may be successful in the market, leading in turn to a definition of the required inputs and the nature and extent of the process required to turn inputs into outputs. Outputs are defined in association with target market definition since a characterization of the buyers or consumers of products and services leads to an understanding of the outputs required. These could be end users, consumers, or other participants in the chain, depending on the step in question. The job or process that the participant plans to carry out within the overall chain, including planned activities and investments, is defined, leading the identification and definition of the inputs required from prior steps in the chain or from external sources, such as other supply chains.

The technology-driven approach would begin with a definition of a new technology, an understanding of the process required to harness the technology and create products and services, and an exploration of the market for suitable opportunities that match the products and services. Having defined the target market, the inputs required to support the process of harnessing the technology would be defined.

6.3.2   Identification of Potential Partners and Collaborators

When defining the inputs and outputs from a step in the chain, it is also possible to identify and describe partners and collaborators in and around the step in the chain. Collaborators within the step in the chain would be involved in the business process being supported inside the step; while collaborators outside the step would be participants in either an upstream or downstream link in the chain. This highlights dependencies related to your role in the chain, the outputs required for the target market, and relationships that need to be developed and managed if you are to succeed. This would also include the investigation of partnering arrangements between public- and private-sector organizations through a sweep of the market and business sectors involved to identify and understand the participants, their resources, capabilities, motivations, business directions, and chosen business approaches.

6.3.3 Identification and Definition of Proposed Business Relationships

Once the participant has defined the step(s) in the chain that are to be addressed and the collaborators involved, the nature of the interaction and the business basis for the relationships can be determined. There is a wide range of potential relationships, both formal and informal, that can be identified, defined, and applied to support your place in the chain and the required relationships with others above and below you in the chain and those within the same link of the chain. These include client-vendor contractual relationships, joint ventures, partnerships, and alliances. The exact form of the business relationship selected will depend on the nature of the parties entering the relationship and the importance of the relationship to both parties.

6.3.4 Proposed Operational Approach

This includes details of the proposed mechanisms for value creation, revenue generation, and profit, including processes to be supported and products and services to be developed or delivered. This may also include the definition of product and service provider to customer relationships, the proposed use of existing distribution channels and relationships, and proposals for increasing, enhancing, or protecting the value of an existing customer relationship and revenue stream. The understanding and effective application of business models in ATIS is one of the primary reasons we introduced in Chapter 5 the concept of value or supply chains with respect to advanced traveler information systems. In order to develop a successful business model, you have to define and understand the value chain within which your business will operate. This requires the development of a frame of reference for your business activities that defines the part your enterprise or organization will play in the overall application of resources to the creation of value. This also identifies, defines, and describes the other participants or potential participants in the value chain and how you plan to interact or compete with them.

From a public perspective, business models and business plans are usually replaced by the development of policy and program plans that identify and describe the transportation policy goals and objectives to be attained and the actions or implementations planned in order to get there. We believe that from a traveler information systems perspective, the policy and program plans are the functional equivalents of the private-sector business models and business plans. In most cases, the public-sector agencies involved in the planning, development, and implementation of advanced traveler information systems will look to such documents for guidance on how to proceed and how to structure investments. One of the limitations of most policy plans is the complete focus on the use of public funding resources to the exclusion of the private-sector equivalents. This seems to stem from a desire to plan and manage only those resources that are under 100% public direction and control.

6.4 Basic Traveler Information Business Models

In Chapter 5 we mapped our operational concept for advanced traveler information systems to the roles and responsibilities proposed for both public and private sectors (see Figure 5.2). We did this by laying out the advanced traveler information supply chain along a vertical axis, and the roles and responsibilities along a horizontal axis with the proposed system architecture configured to show who does what. This is essentially a business model since it defines the process to be supported and the roles and responsibilities of the participants. Obviously, a lot more detail must be added in order to develop a complete business plan that could guide the development and implementation of an ATIS, but it has sufficient detail to enable us to choose between different options and configurations. As you will see later in this chapter, we can use this type of graphical business model to examine and compare business approaches that have been applied to prior advanced traveler information implementations.

This type of graphical model requires that you have at least an outline architecture for the proposed advanced traveler information system in terms of a definition of the major subsystems and the data and communication links between them. Since we believe it imperative that early planning work for traveler information systems (or any system for that matter) focuses on what needs to be done rather than how it should be done and incorporates a strong element of partnership identification and negotiation, we think it is useful to begin with a simpler business model that can evolve over the course of the technical and business-planning activities. This simpler business modeling approach involves a basic mapping of each step in the ATIS supply chain to the public or private sectors. Taking another look at the supply chain we introduced in Chapter 5, we can illustrate the simpler approach identification and definition of business models for traveler information specifically. We can do this at the most basic level by assigning either a public- or private-sector responsibility for each link in the chain. The chain was first introduced in Chapter 5 (see Figure 5.1).

Since there are six links in the chain and two options (public or private), there are a total of 2^6 or 64 theoretically possible public-private combinations of the basic links in the chain. This number grows substantially when you take into account the additional number of hybrids that are possible. For example, we can also decide to allow sharing of any link between both the public and private sectors, with one sector taking a lead role and another taking a subordinate role. This would yield 3^6 or 729 theoretical combinations, and these are just the basic models. The main point we want to make is that there are a large number of options that deserve a thorough review and analysis, taking account of your own organization's particular circumstances and context. The process of analyzing the chain and deciding what your organization wants to do, what it is capable of doing, and with whom it wants to partner are important elements in the development of an effective approach to ATIS.

In terms of providing an illustration of basic business models, Table 6.1 contains a sample of eight basic traveler information system business models that we created by assigning a public or private sector total responsibility for one or more links in the chain.

For illustration purposes, we have defined a completely shared chain, an all-public chain, an all-private chain, and a range of basic models with increasing private-sector involvement and decreasing public-sector involvement.

For ease of reference, we have given each business model option a number and a name. The name reflects the highest point on the supply chain at which the private sector plays a role. For example, model B5 is named private marketing since the highest point at which the private sector plays a role is marketing and everything above is public.

You will note that we have only defined model options with complete public, complete private, complete shared, or a single interface or hand-off between the public and private sectors. Models with multiple hand-offs or

Table 6.1
ATIS Basic Business Models Summary

	All shared—Model B1	All public—Model B2	All private—Model B3	Private delivery—Model B4	Private marketing—Model B5	Private add value—Model B6	Private data fusion—Model B7	Private data collection—Model B8
100: Build the data infrastructure	Both	Public	Private	Public	Public	Public	Public	Public
200: Operate the data-collection system	Both	Public	Private	Public	Public	Public	Public	Private
300: Conduct data fusion	Both	Public	Private	Public	Public	Public	Private	Private
400: Add value to the information	Both	Public	Private	Public	Public	Private	Private	Private
500: Market the information	Both	Public	Private	Public	Private	Private	Private	Private
600: Deliver the information	Both	Public	Private	Private	Private	Private	Private	Private

flip-flopping roles and responsibilities are feasible and can be defined quite easily. We have simply omitted them for clarity since the use of the eight models shown serves our purposes for illustration.

6.4.1 Basic Model B1: All Shared

Each step in the traveler information supply chain is supported and operated by both the public and private sectors to varying degrees. These can be completely independent chains, or may contain interfaces where value is transferred. The most typical scenario in which this model is adopted is described in Sections 6.4.1.1. to 6.4.1.6.

6.4.1.1 Building the Data Infrastructure

The public sector invests in a partial data-collection infrastructure designed primarily to support traffic management. The private sector develops a parallel data-collection infrastructure to collect complementary data. For example, the public data-collection infrastructure may only cover freeways, leaving the private data infrastructure to fill in data collection on arterials and major urban streets. Private data infrastructure could take the form of CCTV cameras or noninvasive infrastructure, such as spotter planes or roving reporters.

6.4.1.2 Operating the Data-Collection System

Both public and private sectors establish and operate their own data-collection systems. There may be an interface where the public sector transfers value to the private sector in the form of publicly collected data. This is often achieved by colocating a private-sector data-collection operator in the public traffic-management center. The operator is given access to the data stream from the public data-collection infrastructure.

6.4.1.3 Fusing the Data

Here again, both public and private sectors establish and operate their own data-fusion systems, taking the raw data and combining different data sources into single, useful data sources. There could be another value interface at this point where the private sector transfers value in the form of fused data back to the public sector, depending on the nature of the agreement allowing the private-sector operator to colocate in the public-sector traffic-management center.

6.4.1.4 Adding Value

The creation of information products and services is also carried out by both sectors. The public sector typically focuses on free information services, while the private sector targets subscriber services or free, advertiser- or sponsor-supported services.

6.4.1.5 Marketing

Here again, there are parallel activities. The public sector focuses on making the consumers and travelers aware of the existence of the traveler information services, the benefits of using them, and the access arrangements. The private sector does all this as well as communicating the cost of using the services or products, placing a higher emphasis on the value proposition (i.e., the benefits to be attained by making use of the offerings).

6.4.1.6   Information Delivery

Two distinct streams of information—one public and one private—emerge at the end of the supply chain. They make use of the same information delivery channels and devices, including Web sites, TV, radio, mobile phones, and pagers. The public sector may also deliver information over public channels, such as dynamic message signs or other publicly owned en route information delivery devices, such as kiosks.

A significant feature of this model lies in the fact that there is no single interface between the public and private sectors and no coherent selection or packaging of roles and responsibilities. This business model can be adopted and implemented with a minimum amount of formal relationship definition between the public and private sectors. The traveler information supply chain can effectively become two parallel chains with optional intertwining at selected points, if desired. This minimizes the effort required to develop relationships and agreements and provides maximum flexibility for each sector.

A significant weakness of this business model lies in the degree of overlap that can develop between public- and private-sector investments. There is no formal supply-chain management required to operate this model, allowing duplicate investments to be made, creating inefficiency through parallel or even conflicting actions on the part of each sector. An example of this would be the public sector delivering traveler information in parallel with the private sector, providing publicly subsidized free information to travelers who may have been prepared to pay for it, thus creating competition. Of course, it could be argued that the public sector should provide the information for free since the taxpayer has already paid to have the data collected and converted into information. This takes us back to one of the basic premises we discussed earlier describing the selection of payment mechanisms according to willingness to pay and the degree of social or community benefit. The public sector, attempting to maximize use of limited funding, tries to find the most cost-effective way to deliver desired services. The lack of structure and agreement in this business model makes it very difficult to take a rational, structured approach to cost minimization, investment optimization, and effectiveness maximization.

6.4.2   Basic Model B2: All Public

As the name states, when this model is adopted, all the steps and activities in the supply chain are implemented and supported by public-sector resources. From the public point of view, this business model has the attraction of offering total control over the operation and execution of all aspects of the traveler information supply chain. This provides the public sector with maximum opportunity to ensure that the transportation policy objectives defined when launching the initiative can be fully attained. On the negative side, this also means that the public sector has to provide all the money required to plan, develop, and operate the entire chain, typically involving duplication of effort and investment being made by private-sector interests.

Note that when we say that the public sector is responsible for all steps and activities in the traveler information supply chain, we are inferring that the public sector provides all the required resources and has full operational and strategic-management control over all such activities. This does not preclude the utilization of private-sector resources within this framework. For example, the public sector could elect to outsource the data-fusion step by developing a scope of work and a request for proposals, seeking competitive bids, then procuring the services of a private-sector data-fusion operator or contractor. In this case the private sector is literally in a work-for-hire situation, with the public sector providing the resources and direction. Any value created in the course of the data-fusion step belongs to the public sector. We will return to the subject of variations on the basic model themes when we discuss practical applications of the models later in this chapter.

6.4.3   Basic Model B3: All Private

This is a complete stand-alone private-sector business model, where all links in the traveler information supply chain are supported entirely by the private sector. One private-sector enterprise or a number of separate private-sector enterprises working in collaboration could carry this out.

This model also features the minimum need for public management of the traveler information supply chain. The public sector takes a hands-off approach, with one or more private-sector enterprises handling all six links in the supply chain. This gives the private sector the maximum degree of freedom, but also the minimum degree of financial or resource support from the public sector.

The advantage of this model to the public sector is the low level of resources required. The public sector hands off the data to the private sector at an early link in the chain and leaves it to the private sector to convert it into information and support all other links in the chain. The disadvantage lies in the lost opportunity for synergy and the duplicate effort and activity that has to be applied. The private sector is likely to be carrying out the same processing in parallel since each enterprise must conduct its own data fusion, presumably to independent standards and specifications. This also intro-duces a quality issue as the information delivered in the same region about the same transportation-network and network conditions will be coming from multiple parallel processes with no mechanism for achieving consis-tency or standardization. Moreover, the private enterprises may practice cream-skimming (i.e., providing traveler information only to the rich urban market and ignoring the relatively poor rural market where traveler informa-tion is still needed). In addition, the private enterprises providing the service may monopolize the market, setting up insurmountable barriers to competi-tors to enter the market.

6.4.4 Basic Model B4: Private Delivery

This model features the minimum degree of private-sector involvement. The public sector extends as far as the penultimate link in the chain, then the pri-vate sector completes the chain by supporting traveler information delivery only. This would require a public-sector infrastructure and resource set capa-ble of producing the required traveler information to the required quality and quantity levels to enable the private sector to effectively conduct busi-ness. In addition, the public sector would be required to carry out all the add-ing value activities, such as information bundling and bulletin creation, as well as support all the marketing functions for the public and private sectors combined. This business model may be appropriate where private-sector information service providers are already targeting the market with informa-tion services and products other than advanced traveler information and wish to add this to the portfolio of information offerings. Under this business sce-nario, the private-sector information service providers would simply add the finished products and services from the public sector to their existing infor-mation delivery channels. The public sector would retain total responsibil-ity for all marketing activities under the auspices of this business model. In practice, this would be highly unlikely to occur, as the private sector would probably have existing marketing arrangements for other items in the infor-mation portfolio.

6.4.5 Basic Model B5: Private Marketing

This approach extends the reach of the private sector from just information delivery as in B4 to include marketing of the products and services also. The public sector handles all links in the chain as far as marketing and provid-ing the data-collection and information-processing facilities to deliver the finished traveler information services and products. The private sector takes over at Step 500: Marketing and supports both the marketing and information delivery steps in the chain. This hands the marketing functions required to make travelers aware of the services, the access arrangements, cost of use, and benefits over to the private sector. The private sector would conduct all marketing for both private and public service deliveries whether they are subscription-based, advertising-subsidized, or free.

6.4.6   Basic Model B6: Private Added Value

This approach transfers to the private sector those activities associated with Step 400: Add Value. The public sector hands off to the private sector at the end of Step 300: Data Fusion. This leaves the private sector with complete responsibility for the downstream end of the supply chain, including the addition of value through the development of service and product packages, the marketing of the products and services, and the delivery of the same. This type of approach would be useful in a situation where the public sector has a focus on transportation management and wishes to outsource the development of more sophisticated traveler information services to the private sector. A deal may be struck to enable the public sector to have access to and use of the private sector outputs from the add-value step.

6.4.7   Basic Model B7: Private Data Fusion

The public sector, having built the data-collection infrastructure, operates it and provides the data to the private sector for subsequent links in the chain. The private sector develops and operates the systems required for fusing the data from different sources, adding value, marketing, and delivery. This effectively outsources the parts of the advanced traveler information supply chain that are not common to transportation management. The most likely scenario for the adoption of this model is where the private sector has a desire to address the market for traveler information and the public sector has an exclusive focus on transportation management. The public sector adopts a hands-off stance and simply provides the data that has been collected for transportation-network-management purposes over to the private sector. The data can be provided for free, in return for an agreement to provide enhanced data and information back to the public sector, or in return for a portion of the revenues that will be derived through the sales of the information.

6.4.8   Basic Model B8: Private Data Collection

This model shows almost complete private-sector operation of the advanced traveler information supply chain. Once the data infrastructure has been planned, designed, and implemented, the public sector hands the data collection and all subsequent activities associated with the supply chain over to the private sector. This could be a useful model in a situation where the public sector has a previously deployed data-collection infrastructure and wishes to outsource future management and operations to the private sector. The private sector would manage and operate the data infrastructure for both transportation management and advanced traveler information purposes.

Exploration of the range of basic business models for advanced traveler information systems is an excellent way to approach the subject in a structured, coherent manner. There are many issues and options that are raised in the course of such an exploration. For example, how does a public agency decide which of the basic models is the most appropriate choice given the need to secure control and yet at the same time make maximum use of private initiatives? We believe that it is very important that both the public and private sectors have a well-thought-out, highly developed business plan as part of the overall approach to the planning, development, implementation, and subsequent operation and management of an advanced traveler information system. The basic models explained here are great starting points for the development of such a plan. We also believe that the exact composition and configuration of the business model will vary depending on specific circumstances, such as whether the organization is in the public or the private sector. Goals and objectives will vary and perspective and cultures will be different. It is common to communicate the nature of the business model to be adopted in the form of a business plan. We will take a look at business-planning approaches for ATIS for both public-sector organizations and private-sector enterprises later in this chapter.

6.5   Enhanced Business Models

The basic models explored in the previous section are a great help in developing an initial approach plan for traveler information systems. As the approach evolves and a greater level of detail is required, however, it becomes necessary to move beyond these simple, relatively abstract models. More practical variations of the basic models described above can be developed by sharing the links in the chain between each sector in various degrees and varying the method of recognizing and rewarding value creation in each step and value transfer between steps. There are many ways in which to reward the creation of value. Direct payment contracts, public-private partnerships, and bartering have all been used to provide recompense in return for reward in the traveler information domain. The relationships required to support public and private activity within the various links of the traveler information supply chain can be defined and detailed. We refer to these more detailed, practical models as enhanced business models.

Rather than just leave you with a collection of high-level theoretical possibilities (although these serve our purpose if they encourage you to think about the possibilities and conduct the analysis work), we decided to describe some of the more detailed hybrid business models and some that have actually been applied in practice. These will illustrate some of the variations on the theme and shed light on the current state of the practice. These models are defined by the relative public- and private-sector roles and responsibilities and the mechanisms used to recognize and reward value creation and transfer.

There are a number of traveler information business models already in use around the world today. They tend to be hybrids or variations on the theoretical models proposed above, reflecting the difference between theory and practice. Differences in the procurement method or in the way value creation is recognized and rewarded lead to variations around the basic themes described earlier. These models were first defined and documented in an excellent report [1] produced by Mark Hallenbeck and his staff at the Washington State Transportation Center. We have drawn on this reference for the names and basic definitions of the models, relating them to our traveler information supply chain and adding our own perspectives on the use and experience gained in the application of the models. The graphics were derived from those contained in [1].

6.5.1 Enhanced Model E1: Public-Centered Operations

Figure 6.1 illustrates this model within the context of the traveler information supply chain.

Note that in addition to mapping the model to the traveler information supply chain, this diagram (and subsequent ones in this section) also provides an additional level of detail by identifying the relationships between organizations used to support the model. Taking a look at Steps 100 and 200, we can see that a total of four organizations are supporting both of these steps. On the public side of the diagram, the transit agency, the state department of transportation, and the city transportation-management system are all

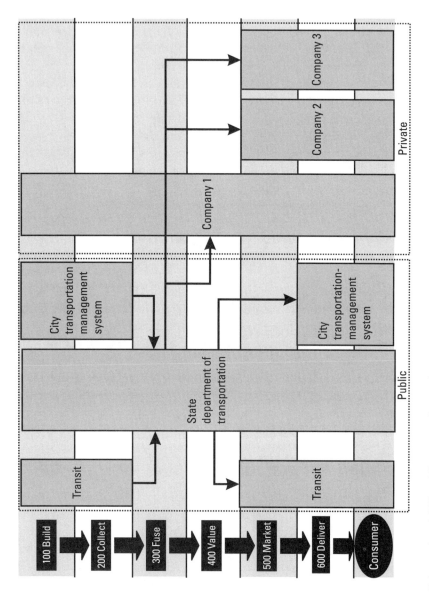

Figure 6.1 Enhanced model E1: public-centered operations.

active in building data-collection infrastructure and operating data-collection systems. This mirrors many practical situations where the total data set required for ATIS can only be obtained by combining data from different jurisdictions or from public agencies with responsibilities for different modes. This is one of the classic features of intelligent transportation systems. The need for cross-organization and interjurisdictional activities is pronounced. We will cover this in more detail in Chapter 7 when we look at how advanced traveler information systems fit into the bigger picture. There is also one private company supporting both of these steps in the chain. This is an information service provider who is addressing all steps in the chain in order to develop and deliver traveler information services and products. For the next two steps there are two organizations providing support: a public-sector agency in the form of the state department of transportation and the private information service provider we have just described. These are joined by the transit agency on the public side and the other two information service providers on the private side for Step 500: Marketing. In the final step all organizations participate when the city transportation-management systems rejoins the chain to provide support for information delivery. The black arrows on the diagram indicate data flowing from one organization to another in the course of system and business operation. The public sector supports all links in the chain, with the private sector joining in on some or all of the links, after data infrastructure and data collection. There is more than one private-sector organization working alongside the public sector, with at least one private-sector entity conducting data infrastructure building and data-collection operations. This gives the public sector a high level of control over the entire chain by involving them in all aspects of traveler information, while providing competition at the marketing and delivery steps in the chain. This business model is based on the assumption that the state department of transportation has the resources and the capabilities to support all steps in the advanced traveler information supply chain. They are designated as the lead public agency in the model. This decision would depend on local and regional factors, and in some situations one of the other public agencies, such as the city or the transit agency could take on this lead role.

The VICS advanced traveler information system implementation as described in Chapter 2 is a good example of the application of this business model. The public sector addresses all the links in the chain with the exception of the last one—delivery—where multiple private-sector enterprises take over, receive the value created, and complete the chain by delivering the information over multiple devices and communication techniques.

An interesting feature of the VICS model is that it is executed on a national basis rather than a regional one. The public-sector operation provides fused, value-added traveler information on a national coverage basis, and the private-sector entities operate information delivery on a national basis also. Another feature of the VICS system is the imbedding of the customers' payment in the equipment cost, thus eliminating monthly subscription, which is objectionable to the users who are used to getting traffic information for free.

6.5.2   Enhanced Model E2: Contracted Operations

Figure 6.2 illustrates the contracted operations business model. This is similar to model E1, with the exception that Step 300: Data Fusion and Step 400: Add Value, while still under public-sector direction and control, have been outsourced to a private-sector company for management and operations. This enables the public sector to make use of private-sector expertise and experience in data fusion and adding value, while maintaining overall control and direction of the activities. This is where the distinction between the public and private roles blurs if only the basic business models are utilized. Another feature of the particular configuration chosen for this model is the activity of company 3. In this case, company 3, a private-sector information service provider, has chosen to support the entire supply chain, while also receiving fused, value-added data from the publicly controlled, privately operated, "contracted fusion and add-value operation." Company 3 has elected to adopt this approach in order to differentiate itself in the market by fusing public data with data sourced from its own private data infrastructure and collection operation, adding more value and developing unique traveler information offerings. This is in contrast with company 1 and company 2, who take the fused and added-value data from the publicly controlled operation and make use of it directly.

The public sector still supports the entire supply chain, with the fusion contractor supplying fused data to both public- and private-sector downstream links in the supply chain. Public and private sectors are responsible for their own add value, marketing, and delivery operations.

6.5.3   Enhanced Model E3: Contracted Fusion with Asset Management

A further adaptation of the contracted fusion and add-value operation featured in model E2 is the addition of an asset-management responsibility to the private contractor's role. This asset-management responsibility includes the development of plans for enhancement of data infrastructure

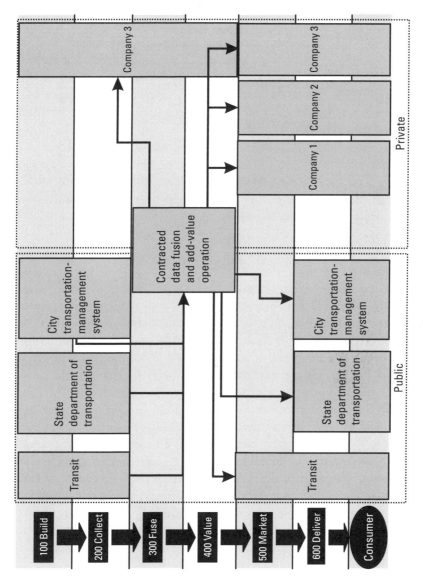

Figure 6.2 Enhanced model E2: contracted operations.

development and data-collection activities, in coordination with market development and business-planning roles. The contractor takes responsibility for a larger component of the public-sector role including the marketing and promotion of the services on behalf of all participating public-sector agencies. Figure 6.3 illustrates this, showing the transit agency withdrawing from Step 500: Marketing and the private contractor migrating down into that step.

The TravInfo® Bay Area deployment introduced and described in Chapter 2 was originally an example of model E1 since the entire traveler information supply chain was addressed by a publicly funded initiative with both federal and local funding for a field and operational test of advanced traveler information systems technologies. More recently, TravInfo® has migrated to this model (E3) with the introduction of a private-sector contractor handling data fusion, adding value, and providing asset-management functions. While still publicly funded, the private sector will conduct additional data infrastructure development and implementation in addition to the data-fusion and asset-management activities.

6.5.4   Enhanced Model E4: Franchise Operations

This model is an adaptation of model E3 and features a transfer of the data-fusion, add-value, and asset-management operations entirely under private-sector direction as well as management and operations. In this model, the public-sector contracts on an exclusive basis with a private-sector enterprise for such services for an entire region. The private-sector enterprise then feeds fused data to the public sector and to other private-sector enterprises as agreed. Figure 6.4 illustrates this model.

The Trafficmaster implementation of ATIS on the U.K. motorway (freeway) and major arterial road networks is an example of this type of business model. Every step in the traveler information supply chain is addressed by the private sector. The building of the data-collection infrastructure on public right of way is enabled by special legislation and a license issued from the public sector to the private sector. This is what we would call a pure-play business model, where one enterprise, Trafficmaster™, addresses all six links in the chain in a self-contained manner. This lends simplicity, clarity, and manageability to the whole chain. Note that the Trafficmaster model, like the VICS business model, is on a national rather than regional basis.

Hopefully the use of these enhanced business models to illustrate possible approaches as well as ones that have been adopted in some selected examples around the world will enable you to identify the issues and some potential approaches. While there is no single universal solution or ideal

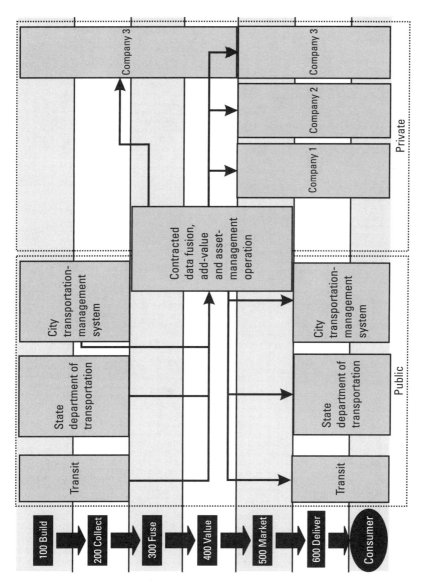

Figure 6.3 Enhanced model E3: contracted fusion with asset management.

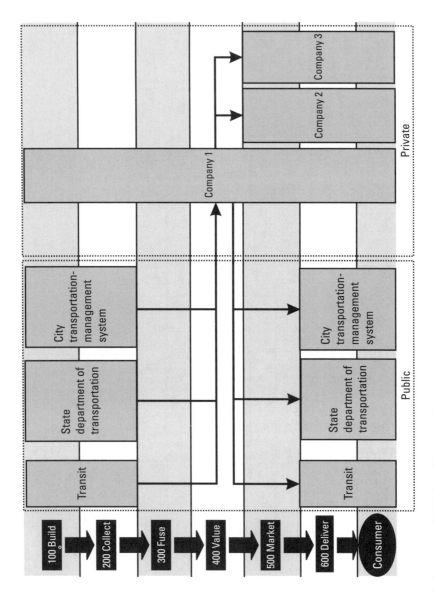

Figure 6.4 Enhanced model E4: franchise operations.

approach, we have shown that there are techniques and tools that can be used to help you to make the best choice.

Reference

[1]   Hallenbeck, M., "Choosing the Route to Traveler Information Systems Deployment", U.S. Department of Transportation, ITS Joint Program Office, Washington State Transportation Center (TRAC), 1998.

7

Business Planning for ATIS

7.1 Learning Objectives

After you have read this chapter you should be able to do the following:

- Define and apply a business-planning and modeling approach for the public sector;
- Define and apply a business-planning and modeling approach for the private sector;
- Recognize the need for public-private collaboration in planning and deployment;
- Understand the importance of a fair and objective determination of the value of traveler information from multiple perspectives.

7.2 Introduction

Having introduced the concepts of basic business models as a tool for initial approach planning and the application of more detailed practical business models or configurations for defining the proposed approach, we now need to pull it all together and complete the business-planning process. The basic and enhanced business models we described in Chapter 6 are the skeletal components of a complete business plan, the essential ingredients required to plan and execute. In order to apply these models in a practical situation, we need an overall approach for selecting the appropriate business model and developing the model definition into a full-fledged business plan or approach. Since the needs and issues of the public sector diverge from those of the private sector as we approach the detailed business planning, we have provided separate treatments of each sector.

7.3 Public-Sector Business Planning

The choice of a suitable and appropriate business model to address public-sector needs, issues, problems, and objectives must start with the identification of requirements and then move to the definition and development of appropriate solutions. The business model must be compatible with policy objectives and program plans that have been previously developed, but must also take account of the private-sector resources that could be harnessed toward the mutually beneficial attainment of public objectives. You can see from the discussion of business models in Chapter 6 that there are a large number of combinations possible. When you include not just the basic traveler information supply chain, but also the variations in how value creation and transfer are recognized, the options available are numerous. The obvious question is how do you choose the traveler information system business model that is appropriate and most effective for your particular situation and needs?

As we have discussed in earlier chapters, it is vital to drive both public- and private-sector involvement in traveler information systems from an objectives-driven perspective. The overall concept should be one of creating a win-win situation in which the public-sector transportation policy objectives are satisfied, while maximizing the leverage possible from private-sector collaborative action and investments. A structure and format for approaching the issues and supporting an effective dialog between both sectors would be extremely helpful. We have a firm belief in the definition of reasonably well-structured approaches to most things relating to ITS and transportation. So it should come as no surprise that we have defined a proposed methodology for identifying and confirming needs, conducting a situation analysis, making a capability assessment of both the public and private sectors and developing the appropriate relationship. Figure 7.1 illustrates our proposed public-sector approach.

7.3.1 Define Public Needs, Issues, Problems, and Objectives

The public sector should identify, define, explore, and confirm the core objectives and motivations that drive the desire to support, invest in, and

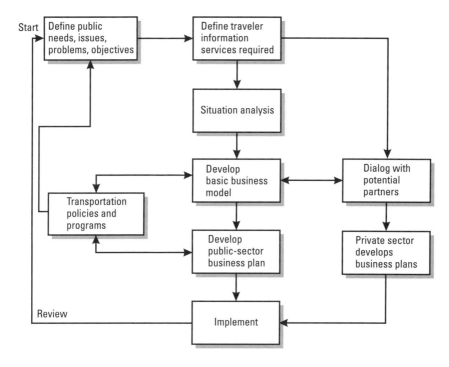

Figure 7.1 Public-sector traveler information business-planning process.

be involved with traveler information systems. The requirements should be identified, explored, and confirmed through a formal system engineering process that enables all decision makers and affected parties to provide input to the development and definition of the requirements to be addressed by the system. The defined objectives should be clearly linked to the community benefits that are the primary drivers in creating the system.

7.3.2   Define Traveler Information Services Required

Based on needs, issues, problems, and objectives, a preliminary set of desired traveler information services can be defined and described. We recommend that these be characterized as what services are desired rather than how to deliver them. The reason for this is that we want to leave the choice of delivery methods and products until later, when a dialog between the public and private sectors can be initiated and utilized to define the required system. The idea is to focus on the end of the traveler information supply chain by considering the range of services that are desired for delivery to the consumer, in full support of the attainment of objectives. Once that has been defined, then you can ascertain the products, systems, and activities required to deliver such services in the most effective manner. The definition of required services should also include a definition of the desired geographic coverage of the services.

7.3.3   Situation Analysis

The situation analysis is a structured assessment of the current status both inside and outside of the public-sector agency. The objective is to obtain a clear picture of current conditions and the capabilities and resources available to be applied to traveler information systems planning and development. This represents a snapshot of the current landscape with respect to traveler information systems. From an internal perspective, there are a number of questions to be answered when conducting this analysis.

- *Market demographics and characteristics:* How many potential customers are to be addressed by the traveler information system? What are their key characteristics; what are their needs and buying preferences?

- *Legacy:* What do you have in place and what does the private sector have in place? What do you have planned with committed investment? If either sector has substantial legacy systems already in place, then there may be no other logical or economic choice than to let the one with the most legacy lead or support that part of the chain. (See [1] for more information on how to deal with legacy.)

- *Legality:* What does the current law allow for various innovative approaches to public-private partnership arrangements? For example, is the public sector allowed by law to share revenue generated from advertising and subscription associated with traveler information? Are cell phones and potentially distractive devices prohibited for distribution of traveler information to drivers?

- *Equity:* How much of the traveler information should be available on an equal or nearly equal basis to all travelers, including the disadvantaged, the handicapped, and those in remote and low-market-demand areas?

- *Synchronizing public investment programs:* There may well be other publicly funded and programmed activities in the transportation area that can provide support and resources at various points in the traveler information supply chain. These need to be identified and coordinated. We will return to this point in Chapter 8 when we discuss the role of traveler information systems in the future big picture for ITS within a region.

- *The bigger ITS picture:* What is the plan for ITS for the region or jurisdiction? How will the traveler information system plan fit with the larger picture?

- *The data requirements for the desired traveler information services:* What data is required in order to support the traveler information systems services that have been defined as required in order to fully meet needs, issues, problems, and objectives?

- *Organizational capabilities and competencies:* What are you good at and how does it match the needs for developing and operating a traveler information system? The number and capabilities of internal public-sector staff may also play a significant role in the decision to use public or private resources at each step in the chain.

- *Financial resources:* What funding is available and what are the terms and conditions?

- *Risk management:* It is not possible to conduct an effective analysis of proposed business approaches without the consideration of risk and its counterpart, rewards or benefits. This should include a review and assessment of the likely risks involved in developing and deploying a traveler information system as well as the probability and likely impact and identification of risk avoidance or mitigation strategies.

The situation analysis must also take account of the wider market outside of the public sector by incorporating a review and assessment of private-sector activities, capabilities, and competencies. This really boils down to who is good at doing what in the supply chain. The capabilities of both public- and private-sector organizations within the supply chain and the region to be covered by the traveler information system need to be defined and evaluated. For example, if the public sector has already planned, designed, and deployed an extensive sensor and communication network in the subject region, then they may be much better equipped and positioned to support the early parts of the supply chain that cover data collection and information processing. This, of course, assumes that the public sector has the appropriate staff resources to operate, manage, and maintain the data infrastructure in support of its original purpose and the desired traveler information services and products. As another example, if the private sector has already established a privately funded traveler information data-collection information-processing and delivery system within the region, then they might be best equipped and positioned to support the data infrastructure and data-collection components of the supply chain.

When the situation analysis is widened to include external markets and participants, another set of questions come into the frame: What products and services are currently available in the market? Which developers and providers are able to meet your needs? It is likely that this activity will also involve a comprehensive review and evaluation of the range of traveler information products and services currently available and a characterization of the provider organizations?

- *Current and future financial viability:* Is the private sector able to sustain a profitable business based on current and likely future market conditions? If this is questionable, then you may lean toward more public-dominated business models.

- *Cost effectiveness:* What is the cost versus performance equation when handing off links in the chain to the private sector? Can they do things more effectively and efficiently or not?

- *Social equity:* This is related to the equity issue mentioned above—how do you ensure that the resultant traveler information system caters to a large cross-section of the population demographic? The private-sector tendency may be to target a small affluent sector of the population in terms of market effectiveness and accelerated returns on investment.

- *Flexible future options:* Is there enough flexibility in the overall approach to allow for market dynamics, changes in requirements, and the introduction of new technologies, products, and services? Does the proposed relationship with the private sector preclude valuable future options?

The supply-chain model we have developed so far can be extremely valuable as a framework for a rigorous assessment of the positioning and capabilities of public- and private-sector organizations. From the public perspective, potential partners in the private sector could be assessed and evaluated against the needs defined for the supply chain. From a private point of view, the capabilities of the public-sector partner to support their part of the over-all supply chain can be explored and examined.

Therefore, we propose that you answer the questions by conducting a traveler information supply-chain analysis. This involves a review of the six links in the chain as previously defined, asking the questions and developing a view on what links in the chain you want to control, which ones for which you want to provide a support role, and which ones you would like to hand off to another entity. Making use of the traveler information supply chain and taking account of the range of available business models, a thorough assessment can be carried out to ascertain the most appropriate and cost-effective public and private roles. The objective is to start at the end of the traveler information supply chain with a good understanding of the services desired, then work back through the steps in the chain, determining the most appropriate roles for the public and private sectors. In simple terms, you should be conducting a structured assessment aimed at answering the funda-mental questions.

7.3.4 Develop Basic Business Model

Once you have conducted the situation analysis, it should be possible, at least at a high-level, to define your proposed business model. We suggest you do that by making use of the traveler information supply chain to describe your preferred business model in terms of links you wish to control, links you wish to support, and links you wish to hand off.

7.3.5 Dialog with Potential Partners

Having developed the high-level business model and equipped with the information from the situation analysis, it should be possible for you to iden-tify and have a meaningful dialog with potential partners in both the public and private sectors. This should lead to revision and realignment of business models and approaches on all sides as the information is exchanged and rela-tive roles, responsibilities, and business directions are explored and con-firmed. This is a good time to confirm and add detail to the information gathered in the course of the situation analysis regarding private-sector capa-bilities, competencies, and business directions.

7.3.6 Develop Private-Sector Business Plan

We put this step in the public approach just to remind you that while you are developing your plans, the private sector may be developing theirs also. The dialog with the private sector is likely to be a mutual learning experience as they will want to understand you and your objectives, capabilities, and competencies as part of their business-planning approach—more about that later.

7.3.7 Develop Public-Sector Business Plan

In our experience, an effective public-sector business plan for ATIS would contain as a minimum the following elements:

1. Objectives statement

    - Soft objectives;
    - Hard objectives;
    - Traveler information service requirements;
    - Community benefits to be achieved.

2. Situation analysis

    - Demographic and market potential analysis;
    - Review of information services market and possible partners;
    - Review of relevant legal and equity issues;
    - Traveler information supply-chain analysis;
    - Traveler information services to be delivered directly;
    - Customer characterization and needs analysis;
    - Market success factors:
        - Data needs analysis;
        - Capabilities and competencies review;
        - Strengths, weaknesses, opportunities, and threats analysis.
    - Review of participants in the information services market;
    - Review of private-sector business models and plans;
    - Review of transportation policies and programs;
    - Basic traveler information systems business model.

3. Strategies and tactics

    - Review, evaluation, and selection of business model options;

- Data sharing policy and plan: define data to be shared and the conditions for sharing;
- Define direct delivery services: what services will be developed and delivered directly;
- Develop strategies for influencing the private sector and other pubic-sector agencies;
- Partnering, procurement, and other relationships.

4. Implementation plan

- Detailed traveler information systems business model with partnering, procurement, and operations assumptions;
- Coordination with public-sector ITS programs;
- Strategies and tactics;
- Investment program;
- Staffing plan;
- Monitoring arrangements;
- Formative evaluation;
- Updating as needed or desired.

5. Executive summary

- Political justification for actions and investment;
- Summary of the key benefits to the region;
- Summary of proposed actions linked to objectives.

7.3.8 Implement the Plan

The next-to-last step in the approach is to implement the plan. It seems obvious, but it is important to actually go ahead and do it, otherwise all the previous effort is wasted. A change of crew may also be required at this point since very often the skills and capabilities required for planning and development are different from those required for effective implementation, operations, and management. A number of crucial factors may also evolve over time, making it necessary to reevaluate roles and responsibilities as the traveler information system evolves and the market for traveler information matures. As technology changes, the range and type of products and services will change also. As the market for traveler information systems evolves and matures, the economic feasibility of some services may be altered. That is why this is not the last step in the approach.

7.3.9   Review the Plan

In recognition of the fact that requirements may change over time and technologies will definitely change over time, it is necessary to define a feedback loop where the results of the ongoing monitoring and management of the plan's performance are returned to the start of the process. For example, it is likely that over the next 10 years or so, there will be an increased availability of in-vehicle information and communication systems. This will enable more traveler information to be delivered to the driver in the vehicle, rather than through the use of en route or roadside technologies, such as dynamic message signs. This may lead to a need to revise the business model, business plans, and investment programs to take full account of the new technology possibilities.

As an overall approach to the development of the business model and the business plan, we would suggest that you first develop a basic mapping of your proposed business model to the traveler information supply chain. This was discussed earlier in this chapter, when we introduced the basic traveler information system business models. Using this as a vehicle for discussion and agreement, address the issues and move towards the development of a more detailed view of the business model, such as those described in the section on enhanced business models. These still have the mapping back to the traveler information supply chain, but they also identify the relationships required in order to support the operation of the model. This will enable you to identify the variations and enhancements to the basic model required to exactly match your situation and context. You can move forward from this point and define exactly how you plan to support each of the identified relationships in terms of agreements, contracts, or partnerships.

7.4   Private-Sector Business Planning for ATIS

The description and discussion with respect to the public perspective on business models for traveler information systems is also applicable to the private sector in many respects. In terms of approaches to the development of business plans and the options available towards the end of the traveler information supply chain, however, the private sector has additional opportunities and challenges.

In order to illustrate the way in which your proposed business approach will generate value for your customers, it is first necessary to understand your customer's needs. As discussed in Chapter 2, the user base or customer group for traveler information is not homogeneous, and customer needs are dynamic depending on mode of travel and current stage in the journey life cycle. Important private-sector questions include the following:

- Who is the customer?

- What is the desired position for private enterprises in the value chain?

- Who are the public and private partners?

- Who are the current and potential competitors?

- What are the competitive advantages of various partners and competitors?

- What is the operational approach?

- How is opportunity converted into value?

- What are the existing and required relationships between the product and service provider and the intended customer?

- How do you make full use of existing distribution channels and customer relationships?

- How do you increase, enhance, or protect the value of existing customer relationships and revenue streams?

Figure 7.2 illustrates the typical approach that the private sector might take toward the development of a business plan and the selection of a business model for ATIS.

7.4.1 Define Business Objectives

These should include both hard and soft business objectives: hard in the sense that the goals would be objective, measurable, and clearly stated—probably in terms of business performance or financial parameters; soft in the sense that these goals would be more general and subjective—described in terms of overall impacts and effects.

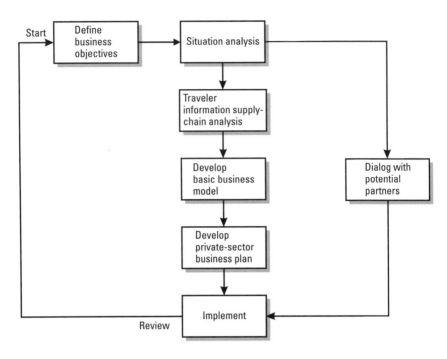

Figure 7.2 Private-sector ATIS business-planning process.

7.4.2 Situation Analysis

A situation analysis comparing internal competencies and capabilities with market features and success factors should be conducted. This includes a determination of the enterprise's current financial, marketing, and technological capabilities and competencies in relation to competitors and those required for market success. The market to be addressed should be explored and defined in terms of the overall market size, the various market segments that comprise the overall market, the nature and characteristics of the customers that reside in the market, and the market segments and the overall needs and wants of these customers. Target market shares for each segment should also be defined at this point.

7.4.3 Dialog with Potential Partners

Having an overview of the traveler information systems market and an understanding of internal competencies and capabilities, a dialog should be opened with potential partners in both the public and private sectors. Positioning in the value chain with respect to these partners and potential relationships built around proposed business models should be explored at this stage of the planning process. While proprietary information may be protected through such means as nondisclosure agreements (NDAs) in dialogs with potential private partners, the same would be difficult to achieve in general with potential public partners. The timing and sequencing of discussions with potential private and public partners are therefore strategically important.

7.4.4 Traveler Information Supply-Chain Analysis

This is the same process we have defined for the public sector, but now approached from the perspective of the private-sector enterprise: define the competencies and capabilities required to successfully address each link in the traveler information supply chain within the context of the target regional or national market, identify the desired roles and responsibilities within the chain, develop a first approximation of the proposed business model, etc.

7.4.5 Develop Basic Business Model

This should be the highest-level business model, defining who does what in the six links of the traveler information supply chain. The goal is to develop a skeleton basis for the detailed planning and decision making required to define a comprehensive detailed business plan for ATIS.

7.4.6 Develop Private-Sector Business Plan

The culmination of the business-planning process is the development and agreement of a detailed business plan that defines the objectives, current situation, desired situation, and the strategies and actions necessary to realize intended goals. In our experience, an effective private-sector business plan for ATIS should contain as a minimum the following elements.

1. Market overview

   - *What you are selling:* description of the proposed service or product offering;
   - *Market size:* sales and revenue forecast for the market and key segments;
   - *Customer characterization and needs analysis:* customer needs and buying habits.

2. Objectives statement

- *Soft objectives:* subjective, nonfinancial, and qualitative;
- *Hard objectives:* objective, financial, and quantitative;
- *Vision of desired position in the market:* more detailed description of the proposed market position including position in the supply chain and relationships.

3. Situation analysis

- *Competition profiles:* development of a detailed profile of major competitors including their resources, capabilities, and perceived business models and market direction;
- *Market success factors:* identification and definition of the most important factors for success in the market and market segments to be targeted;
- *Strengths, weaknesses, opportunities, and threats analysis:* detailed analysis of data from the market overview and the creation of a structured view of the enterprise's current position;
- *Competitive position:* review of enterprise resources and capabilities contrasted against the major competition identified through market surveys.

4. Strategies and tactics

- *Strategies:* plans for addressing objectives and points from the situation analysis. For example, if the market for traffic information alone is insufficient for the private enterprise to be profitable, the strategy would be to bundle the traffic information with other information of interest to travelers, such as news, weather, stocks, and sports. This in turn would lead to strategic alliance with public or private partners that can provide such information in profitable terms.
- *Tactics:* specific actions and activities designed to apply the strategies with associated cost estimates. For example, a private partner that is successful in one region may enlist the help of its public partner in that region in publicizing their success with the objective to promote the private partner delivering similar services using a similar pattern of public-private partnership arrangements in other regions.

5. Financial plan

   - *Investment needs:* cash flow profile for the duration of the plan;

   - *Sales, revenue:* sales volumes, anticipated unit prices, and fore-casts for revenues;

   - *Net income forecasts:* illustration of the plan for making money, taking account of the investment needs, sales and revenue fore-casts, and resource requirements.

6. Resource plan

   - *People resources:* details of the types of staff and management structure required to successfully implement the plan;

   - *Other resources:* details of the other things needed to develop and sustain the business including office space, hardware, software, data, and other major purchases foreseen.

7. Monitoring arrangements

   - The arrangements for measuring and monitoring progress in implementing the plan;

   - Definition of performance measurements;

   - Mechanisms for review.

8. Executive summary

   - Overview of the whole plan designed for review by key decision makers and investors.

There is a great deal of interdependence between the various elements in the outline. This makes it virtually impossible to address the development of the business plan in a linear-sequential manner. There will be significant cycling back and forth between elements until convergence on every element in the business plan is achieved. You will note that there are similarities between the content of this private-sector business plan outline and the one suggested for the public sector. This is not surprising, since both sectors are addressing similar markets and objectives. The main differences lie in adaptations of the outline designed to address diverging public- and private-sector needs and cultures. For example, when setting objectives the public sector must give a lot more attention to existing transportation policies and programs. On the other hand, the private sector tends to invest more time in assessing and evaluating the competitive landscape within which the business will operate.

7.4.7   Implement the Plan

As was the case with the public-sector plan, the real value is only unlocked when the plan is adopted and implemented. In this case the plan would be used to seek and acquire investment funding, guide the development of the business and organization required to support the plan, and support dialog and collaboration with the public sector.

7.4.8   Review the Plan

The plan should be subject to regular reviews, linked to business monitoring and progress reviews. These are typically carried out on a quarterly basis. In some particularly fast moving business environments, however, reviews may be held at much shorter intervals. Many private-sector start-up enterprises measure progress in terms of the burn rate. This is the speed at which they are spending investment money to set up the business, position, and prepare for a return on investment. If the burn rate (in, say, dollars per week) is high, then weekly progress reviews may be appropriate. On these occasions, progress against the plan should be checked and actions defined to correct any deviations. The overall relevance of the business plan should also be checked at regular intervals to account for the dynamic nature of information and communication technologies and the rapidly changing market for information services and products.

7.5   Downstream Business Models for the Private Sector

Looking at the second half of the traveler information supply chain, there are a number of opportunities available to the private sectors, relating to the wider context of information services and e-commerce. You will recall when we discussed the traveler information supply chain within the wider context in Chapter 5, we noted that the end of the chain was located in a much wider market for information services reaching beyond the transportation domain. This wider context offers much more flexibility and a higher number of opportunities for the private sector to develop and apply business models.

Most of these opportunities are associated with the add-value, market, and deliver links in the chain.

We carried out some research into the different business models being applied in this wider world of information services and were very fortunate to find a structured list of business models that had been developed by Professor Michael Rappa at North Carolina State University [2]. This work was aimed at documenting and structuring the range of business models that are currently being applied in the world of e-commerce and Web business. We believe, however, that many of the models identified and explained are directly relevant to the ATIS context. Since the use of mobile and fixed Internet technologies represents a significant part of the multichannel information delivery capabilities utilized in traveler information systems, it makes sense that such models would be applicable. Taking this as our basis, we have identified the following business models that might be applied to ATISs.

Advertising Model

This model can be applied through the use of traditional broadcasting media, such as TV and radio, or through more innovative approaches, such as mobile or fixed Internet. The same principles apply in that the information is delivered to the user free of charge, with revenue generated by advertisements and sponsorship messages embedded in the information stream. With respect to traveler information systems, this could involve the delivery of traffic bulletins on TV or radio, sponsored by advertisers, or the delivery of free information on a Web page that also features advertising. A variation on this model would feature the delivery of information at a reduced cost rather than or free. The subsidy required from advertisers would consequently be lower.

Free Model

The users of the information delivery channel are offered something for free in return for making use of the channel. This is intended to generate a high volume of usage, enabling the delivery channel operator to attract advertisers and sponsors to pay for advertisements to be delivered on the delivery channel. With respect to traveler information systems, it may unfortunately be the case that the free item is traveler information. This could be offered as an enticement for the user to visit a Web site regularly, or tune in to a broadcast delivery channel on a regular basis. Another alternative is that the traveler information is the saleable product, with other things being offered for free as an enticement for the user to sign up for paid traveler information.

Subscription Model

In this model, the user of the information delivery channel pays a regular subscription in return for access to the channel. The subscription fee is typically a monthly or annual fee, in return for which the user is given access to the channel through a unique user name and password or other access control method. This model relies on the user perceiving sufficient value in having access to the information to justify the investment in the subscription.

Utility Model

This model is so named because it is the model that utility companies (gas, electricity, water, and telephone) use. It is a variation of the subscription model with pay-as-you-go charging rather than a flat regular subscription. The user is charged according to the amount of usage of the channel, rather than for access to the channel for unlimited use over a specified time period. This allows differential charging according to the value of the information, the timing of access to the information, and the method of accessing the information. For example, traveler information could be supplied to a user on a fee-per-bulletin basis, or a fee-per-second or -minute of access time. The use of an interactive voice response system could be charged on the basis of a premium added to the basic cost of the telephone call. Alternatively, traffic alerts on customized routes during certain hours could be provided for a monthly fee. The ability to process a high volume of small-value transactions in an efficient and effective way is an important element in the operation of this business model.

Affiliates Model

In this model, other service providers are given a financial incentive to encourage their users to make use of the target information delivery channel. This can be achieved by offering a link from the affiliate information service to the target service. Currently, the best way to achieve this is to make use of Internet technologies that enable users to jump to the target system, while those affiliates that provide the Internet traffic are recorded for later payment of the financial incentive. The use of this model harnesses a wider range of information delivery channels to funnel users and potential users to the target information delivery channel.

Portals Model

There is actually a family of models that fit within the overall portals category. The overall concept of a portal is to provide an information delivery channel or convenient source for information and access to services that is so valuable and convenient to users that they make use of it on a regular basis. The portal is the recognized door or entry point for access to the information and services required. The entry point becomes a source of revenue from advertising targeted on the users of the portal as they pass through to the information and services destination they seek. Well-known examples of Web-based portals include Yahoo, Lycos, and Metacrawler, all of which offer access to a wide range of information and services with advertising targeted to the nature of the request made by the user. For example, if you go to Yahoo and search for traveler information, the advertisements displayed on the results page would be selected in terms of relevance to travel, traveler information, or a closely related subject.

Portals can be categorized as generalized portals, personalized portals, or specialized portals. Generalized portals, such as Yahoo, offer a broad and diverse range of information and services, rather like a department store. Personalized portals enable the user to specify the content, format, and structure of the information delivery, tailoring it to personal tastes and needs and enabling the user to take greater ownership through developing the channel. Specialized portals are targeted on specific customer groups and carry information and services that are considered to be of interest to that group. In the realm of traveler information systems, the recent dedication of the telephone number 511 for national traveler information access in the United States is a good example of the establishment and development of a specialized portal. The intention is that on a national basis, users wishing traveler information will be able to dial 511 and be given access to interactive voice-response-driven local traveler information. The 511 portal could also be operated as a specialized portal if the user was able to customize the delivery of the information to specific requirements and preferences.

Here again, our intention has not been to create a comprehensive and exhaustive catalog of business options, but to provide you with some serious food for thought as you develop your own approach to business planning for ATIS. Most business-planning approaches will have to comply with the specific requirements of the enterprise or the venture capitalists, so it is not possible for us to develop a detailed recipe for the business plan. We have provided the distilled essence of a business plan by explaining the generic content and form extracted from a number of business plans we have either created or reviewed. This should enable you to make a start towards a customized plan for your enterprise.

7.6   The Value of Traveler Information

Although we have provided a separate treatment for the public and private sectors with respect to the development of business plans for ATIS and markets, there is substantial overlap, potential for synergy, and common ground between the two. We have recognized this by including elements in both outlines that accommodate dialog and coordination between both sectors. When we took stock of the possibilities for common ground and dialog between the sectors, one dimension stood out as the single most important: determining the value for traveler information. On the public side, the value of traveler information is required in order to justify the public expenditure and show the link between investment and community benefits. From the private perspective, the value of traveler information is a key assumption in the development of business models and revenue forecasts. We believe that the value of traveler information is one of the most important parameters affecting the attitude, positioning, and investment abilities of both the public and private sectors in the traveler information systems domain. Throughout this book, we have advocated a specific approach to the planning design and development of ATIS. We believe that it is essential to the success of such efforts that transportation policy objectives and end user needs drive activities. To do so successfully, to develop effective partnerships that make best use of public and private resources, an accurate and fair determination of the value of traveler information is required.

We believe that this is an aspect of ATIS that requires further research and investigation. The application of mathematical simulation models for specific regional and city networks, combined with the application of decision theory and value-analysis techniques as utilized in other disciplines, could yield some important tools for the measurement and assessment of the value of traveler information in specific circumstances. This would provide the basis for the fair and independent estimation of the value of traveler information that is required as the foundation for a rational approach to the development and operation of ATIS. With ITS applications focused on user services, the value of such services to the end users (namely, the travelers) is supposed to be the basis for investment and operational decisions by both public and private players. Among the 400 to 500 ITS operational tests conducted around the world, most have included the evaluation of benefits of the tested ITS services to the users. Practically all the evaluation, however, has been conducted on an aggregate basis, providing measurements in terms of percent reduction in travel delays or increase in throughput [3]. Moreover, the benefits are usually given in a wide range for a class of user services—for example, up to 20% reduction of travel delays by rerouting traffic through variable message signs (VMS) [3, Table 3.2].

To help ATMS and ATIS investment and operational decisions, one needs a solid basis for assessing the value of traffic information (VoTI) to the end users (the customers) in specific situations so that the public and private players can answer the kind of questions listed below with confidence:

- With a limited budget for installing additional traffic detectors, where should they be located within the road network?

- Can the cost of installing another $200,000 variable message sign at a major intersection be justified based on the benefit to the travelers at that intersection?

- If additional traffic detectors are installed by a private traffic information provider, could it reasonably expect to recover the investment and maintenance costs through additional subscription?

- For an ATIS public-private partnership, which portion of the traffic information should be provided by the public partner based on cost-benefit analysis and which by the private partner with a profit motive?

- How much is traffic information alone really worth to the end users so that the strategy of bundling it with other information (yellow pages, stock quotes) can be developed for telematics to provide information to people on the move?

While these are very important practical questions, there have been no good answers. The question of VoTI was raised in a couple of work sessions in 1999 associated with the AZTech project, which involved a fairly large cross section of seasoned ATMS and ATIS practitioners from both public and private sectors in the United States [4]. Few clues were offered on how to seek good answers. One not-so-satisfactory answer was to wait for the answer in the evolving market. Yet everyone knew that, in the real world, decisions had been made that imply answers, rightly or wrongly, to these questions. It is clear that assessment of VoTI is an area in which significant development work is required in order to provide both a theoretical underpinning and a way to tap the experience of relevant decisions made in the real world.

Our proposed theoretical approach to the determination of the value of traveler information is based on the discipline of decision analysis. Although decision analysis has been well established as a discipline for over 30 years

[5], it has not been applied very much to the field of ITS. One of the authors of this book has introduced a more advanced extension of the discipline to the ITS community [6] to illustrate how win-win solutions could be found for ITS public-private partnerships using social decision analysis. Decision analysis provides a systematic framework for rational decision under uncertainty. Within this framework, the value of information can be determined quantitatively as the difference between the expected values of optimum choice of alternatives with and without the information.

References

[1] McQueen, J., and B. McQueen, *ITS Architectures*, Norwood, MA: Artech House, 1999.

[2] Rappa, M., North Carolina State University, http://digitalenterprise.org/models/models.html.

[3] Chen, K., and J. C. Miles (eds.), *ITS Handbook 2000*, Norwood, MA: Artech House, 1999.

[4] Chen, K., and R. McQueen, AZTech TTN Project Report, 34, Orlando, FL: PBS&J, 2000.

[5] Raiffa, H., *Decision Analysis: Introductory Lectures on Choices Under Uncertainty*, Reading, MA: Addison-Wesley, 1968.

[6] Chen, K., and T. B. Reed, "Social Decision Analysis for IVHS," Proceedings of the IVHS America 1993 Annual Meeting, 1993, pp. 12–19.

8

How Traveler Information Systems Fit into the Future Big Picture

8.1 Learning Objectives

After you have read this chapter you should be able to do the following:

- Define the range of possibilities for sharing and synergy with other systems and initiatives from a public-sector perspective;
- Define the relationships between traveler information and other ITS applications as described in the U.S. National Architecture for ITS;
- Define the range of possibilities for sharing and synergy with other systems and initiatives from a private-sector perspective;
- Describe the use of traveler information as soft traffic management;
- Describe the potential for sharing sensor data between traveler information and traffic-management applications;
- Understand the potential for public-private collaboration;
- Define the potential future convergance of traveler information, location, and payment-systems applications.

8.2 Introduction

This chapter provides some context for the information and advice on traveler information systems we have provided thus far. One of the central tenets in our practical approach to the planning, development, design, and operation of such information systems has been the maximization of synergy and the aggressive pursuit of integration and sharing. The crosscutting nature of information and communication technologies and the inherent need to share data and information for success in the application and operation of ITS makes this the only rational course of action. The exact nature of the integration and collaboration will, of course, vary considerably from one application or project to another based on the nature of your organization or enterprise. For example, some transit information systems have lost-and-found subsystems or functions integrated with traveler information services as the same department supports both services to the public. Private-sector traveler information implementations may bundle other information services together as real-time, perishable information streams. We will not even start to list all the possible connections, interfaces, and extensions that may be possible as traveler information systems are developed and deployed within specific regional and organizational settings. What we think is of much more value across a wider spectrum of readers, and what we want to provide in this chapter, are a few thoughts on the major possibilities for sharing and synergy with other systems and initiatives as traveler information systems are deployed and operated. As in previous chapters, we will provide both public- and private-sector perspectives on the subject, as they tend to be so very different and shed light on the subject from diverse perspectives.

8.3   Public-Sector Perspective

From a public-sector perspective, sharing, synergy, coordination, partnership, and collaboration are common threads that run through the world of intelligent transportation systems. Information and communication technologies support data and information sharing. They work best in a holistic network approach in which the investment in technologies is shared by many applications and multiple organizations. Take high-speed, high-capacity telecommunications as an example. If you decide that the best way to satisfy your need for such communications facilities is to develop, install and operate a fiber-optic communication network, you will need a lot of money. You will have to install the conduits and the fiber, install the communications electronics, and manage and operate the system, just like you would a highway. However, like many information and communication technologies, if you can find someone that has already invested in their own network and is prepared to let you use it for an appropriate price, you will probably require a lot less money. This is because many information and communication technologies have relatively high fixed or initial costs (known in the highway building business as capital investment) and relatively low marginal or incremental costs for expansion. In simple terms, it is not much more expensive to install a 142-strand fiber-optic cable than it is to install a 78-strand one. Also, it is very difficult to accurately predict current and future telecommunication needs, so most people tend to install more than they need, opening up the possibility of renting or selling the reserve or spare capacity. The same sort of mechanism can work for sensors and information processing as well. If one agency acquires and installs traffic sensors along the highway network, then that data could be made available to another agency that needs the same data for a marginal cost, less than the initial cost of developing and installing the sensor network. With respect to information processing, an agency's computer could run additional processes for another agency at a relatively small cost.

The application of information and communication technologies to transportation seems to lend itself naturally to the big picture view of an enterprise's activities and initiatives. In the United States, Europe, Japan, and other countries, this has been recognized through the development and application of national system architectures, or future big pictures that illustrate a comprehensive end game or desired future state. They define all of the major subsystems and interfaces required for a coherent, complete system that addresses previously identified, defined, and agreed-upon requirements. Some also offer an incremental implementation strategy to get you from where you are today to where you want to be tomorrow. Such architectures are a great tool for understanding the wider context within which traveler information systems will operate, offering the possibility to see and comprehend the collaboration and integration that may be possible. Because this is not a book on ITS architectures [1], we will not spend time reviewing the various national architecture approaches. Instead, we will use the U.S. National Architecture for ITS as an example. This is not a better architecture than others developed around the world, but we are much more familiar with this one. All three of us took part in the development of the National Architecture for ITS in various roles. If you would like to learn more about the National Architecture then see [2] for more information. The U.S. Department of Transportation has done an excellent job of documenting the National Architecture and making the information as accessible as possible. We will look at this architecture from a number of perspectives in order to illustrate the importance of ATIS within the overall deployment or ITS. We show how the use of a future big picture in planning the development of a traveler information system can be valuable. Finally, we describe and explain the traveler information components of the future big picture and indicate the relative position of traveler information components within the overall framework.

8.3.1 Traveler Information Within the Context of the U.S. National Architecture for ITS

The U.S. National Architecture for ITS (henceforth referred to as the National Architecture) provides a high-level framework for planning, developing, and deploying ITS in a coherent systematic manner that is driven by user needs. In the course of the development of the National Architecture, a number of system engineering terms entered the vocabulary of the transportation profession. It is worth introducing these and providing a brief explanation before we go any further.

8.3.1.1   ITS User Services

In order to capture the needs, issues, problems, and objectives to be addressed by the big future picture or architecture, system engineers working on the development of the National Architecture made use of the user services concept. A user service is simply a concise, plain-language description of what the eventual system has to do or provide in order to satisfy the needs, issues, problems, and objectives of all defined users of the system. These are the guiding requirements of system development, and they provide an effective mechanism for ensuring that the needs of the end user are fully understood and are used to drive the development of the system. You can think of their usefulness in terms of feedback and feed-forward. The former means the use of the user-service definition to confirm that user needs have been understood and properly articulated. The latter means use of the user services to guide the work of the system engineering and development team in defining the required system component interfaces and operating concepts.

8.3.1.2   Logical Architecture

A logical architecture is a technology independent view of the whole system required to satisfy all of the requirements. In order to maintain this independence from technological solutions, the framework is developed by focusing on the definition of the data that needs to be fed into the system, the processing required, the data flowing around the system, and the data that should come out. This helps to identify where data is being utilized so that the system can be configured, as far as possible, to collect input data once and make use of it many times. It also enables the identification of opportunities for bringing together common processing or work activities, supporting specialization, the use of special processing tools, and maximum efficiency.

For the purposes of logical architecture development, a data flow is defined as a description of the data—in terms of content, format, and structure—that needs to be supplied to the system, that flows around the system, or that comes out of the system as a result of the processing carried out by the system. Here is a small sample of the data-flow descriptions for data that flows in and out of the Driver and Traveler Services subsystem of the National Architecture.

Data flows are usually described in two ways: with short labels and with a brief description. The descriptions are collected together in a data dictionary that can be used to look up particular data flows.

Another concept used in the development and definition of the logical architecture is the process. This is a plain language description of the work, activities, or processing required to convert data coming into the system or subsystems into data going out. Like data flows, processes are described by two components: brief process labels and a process description known as a process specification (Pspec).

As an illustration, here is an overview of the Pspec for Process 6.5.1 "Collect and Update Traveler Information" from the U.S. Department of Transportation's National ITS Architecture version 3.0:

> This process shall collect and update data about incidents, road construction, weather, events and yellow pages data. This data shall be obtained by the process from other ITS functions and from outside sources, such as the weather service, yellow pages service providers and the media. The process shall load the data into a local store for use by the process that provides yellow pages information and reservations.

Data flows and processes are mapped graphically in data flow diagrams. These show the flow of data into the system, from one process to another, and out of the system, explaining how it all fits together and where the processing takes place. These are often known as bubble charts since the processes are represented as circles or bubbles and data flows are represented as arrows. Figure 8.1, an extract from the National Architecture for ITS Logical Architecture document, shows, at a high level, the logical architecture view of the U.S. architecture. It is not possible to see all the detail on this version of the chart, so we have highlighted the high level of interconnectivity between

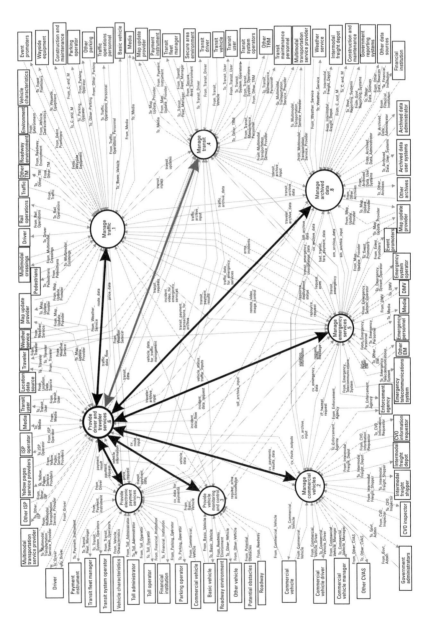

Figure 8.1 High-level logical architecture from the U.S. National Architecture for ITS.

the eight major processes defined in the architecture, again emphasizing the need to take account of partnership and synergy when planning, developing, and operating a traveler information system.

Since the bubbles on the data-flow diagram represent processing, and it is logical to assume that a single organization or enterprise will carry out all processing within a bubble (otherwise it should be split into two separate subbubbles), then the logical architecture can also be viewed as an organizational or partnership blueprint for the traveler information system and its relationship to the wider world of ITS.

The bubble charts from the National Architecture, in our opinion, represent the first level of detail required for the development of a value chain or business process for the operation and management of ITS. The processes and data flows provide a great initial picture of who does what and where in the overall organizational framework.

8.3.1.3 Physical Architecture

Building on the shape of the system as defined by the logical architecture, we can add technology dependence and create what is known as a physical architecture. Figure 8.2 shows the highest-level physical architecture diagram for the National Architecture.

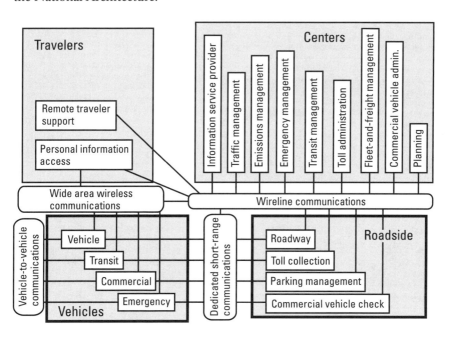

Figure 8.2 High-level physical architecture from U.S. National Architecture for ITS.

This adds technology dependence, but not specificity, since basic technology choices, such as the use of wireless as opposed to wireline communication technologies are made without detailing the exact technology. For example, if wireless is selected as the technology (as in the interfaces from the vehicles subsystems to the roadside subsystems), then multiple options for detailed design and specification of technologies are still open. Various radio communication technologies, such as microwave are possibilities as are wireless communication technologies utilizing laser or infrared light.

Note that this physical architecture view provides us with another perspective on the relationship between traveler information systems (labeled information service provider in this diagram) and other subsystems within the overall big future picture.

This particular architecture assumes that there are shared wireline communications facilities linking the traveler information elements of the system with other management centers, such as traffic-management and transit-management centers. It also assumes the shared use of roadside subsystems, such as traffic sensors.

8.3.2   ITS User Services Addressed

Let us now take a look at traveler information within the context of the national architecture, from an ITS user service perspective.

We conducted a brief analysis of the architecture documentation and identified the three primary subsystems that relate to ATIS:

1. Personal information access;
2. Remote traveler support;
3. Information service provider.

These are highlighted in Figure 8.3, which shows the physical architecture from the U.S. National Architecture for ITS, with the personal information access, remote traveler support, and information service provider subsystems highlighted. Note that the personal information access subsystem supports the use of mobile internet and handheld devices to obtain and deliver ATIS using wireless communication links, and that the remote traveler support subsystem encompasses the use of similar devices but makes use of fixed or wireline communication facilities. A good example would be the use of an Internet connection at home to gain access to traveler information services. The information service provider subsystem covers the provision of traveler

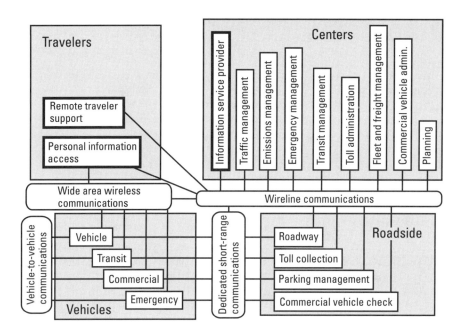

Figure 8.3 High-level physical architecture from U.S. National Architecture for ITS with traveler information systems highlighted.

information services to travelers through the other two subsystems, making use of data collected from the vehicle and roadside subsystems. Looking at these three subsystems, we traced our way back through the architecture to discover which ITS user services were addressed by these systems. We noted that 1,254 of the ITS user-service requirements (USRs) defined for the National Architecture for ITS are addressed by these three traveler information subsystems. The list amounts to 25 pages of text, so we will not reproduce the whole thing here. Table 8.1 shows a selection of the ITS USRs related to traveler information systems to give you an idea of what services are actually supported by the three traveler information subsystems.

The most important point to note is the large overlap between traveler information user services and other needs, issues, problems, and objectives. It turns out that traveler information is one of those core subsystems that have the potential to act as a sort of pivot point for the whole system. Addressing the requirements associated with traveler information systems will also address at least some of the requirements associated with other subsystems. Here again, we see clear evidence that the development and operation of a traveler information system can be most effective when treated as an

Table 8.1
Selection of USRs Addressed by ATISs

User Service Requirement	Name	Subsystem
1.0	Inform Traveler	Remote traveler support
1.0	Provide Traveler Emergency Message Interface	Personal information access
1.0	Provide Personal Portable Device Guidance Interface	Personal information access
1.0	Provide Personal Portable Device Autonomous Guidance	Personal information access
1.0	Provide Personal Portable Device Dynamic Guidance	Personal information access
1.0	Determine Personal Portable Device Guidance Method	Personal information access
1.0	Provide Traveler Personal Interface	Personal information access
1.0	Get Traveler Personal Request	Personal information access
1.0	Provide Traveler with Personal Travel Information	Personal information access
1.0	Collect Price Data for ITS Use	Information service provider

element of a larger system, or as one component in a larger portfolio of initiatives.

8.3.3   Possibilities for Synergy

In order to bring the concepts and approaches described in the previous section to life, we will explore some specific synergies that are possible and can be identified in the National Architecture. Taking the traveler information system—or subsystem, depending on how wide your vision is—as the central focus, we can define the following specific opportunities as examples of potential data sharing between each of the other subsystems.

8.3.3.1   Traveler Information as Soft Traffic Management

When exploring the synergy between traveler information and traffic management, the use of traveler information as a traffic-management tool becomes an interesting concept. Considering the options open to the traffic and transportation professional when it comes to managing the transportation network as efficiently and effectively as possible, there are two major approaches available: provide transportation capacity to meet the demand, or attempt to modify or manage the demand to match the available capacity. Since ITS has a great deal of relevance to the operation of existing infrastructure, many applications relate to the latter category of demand-management implementations. Within the demand-management category, the range of tools or approaches can be defined as either soft or hard.

Hard Measures

These measures have a firm, direct, enforceable impact on the operation of the transportation network. The use of one-way streets, banned turns, and height, width, speed, and weight restrictions to manage the flow of traffic on road networks are examples of these types of measures. Restricting the boarding of buses to specific locations (bus stops) and having railway stations where passengers can only alight are good examples of hard measures from the world of public transit. The psychology at work here can be characterized as "you must" since the measures are usually enforced by law or physical arrangements. The traveler is given no choice and is instructed on how to make best use of the transportation infrastructure.

Soft Measures

Soft measures have an indirect effect and involve the use of influence or persuasion rather than rules, regulations, and mandatory requirements. These measures apply pressure on the traveler toward a specific course of action or behavior, but ultimately leave the choice to the traveler. The kind of mechanisms that can be employed under this category include informational persuasion and fiscal persuasion—that is, providing information that is aimed at travel behavior change or altering the cost of making a journey to influence travel patterns. The psychology can be characterized as "you should" since powerful advice is given but no mandatory instructions or rules are enforced.

With regard to the first area, the use of traveler information systems may hold out the possibility for more accurate, more flexible, and more relevant traffic management. Many of the traditional approaches incorporate permanent infrastructure, such as road signs and marks that convey a rule or instruction that is determined to be the most efficient on some sort of average basis. Given the extremely dynamic nature of traffic flows and usage patterns on transportation networks (consider the difference between peak and off-peak use), the application of traveler information systems as we have defined them could provide some valuable results. For example, a restricted right turn at an intersection could be applied on a part-time basis, rather than 24 hours per day, making use of information delivery techniques, such as dynamic message signs or in-vehicle information devices to inform the driver. Traffic lanes could also be assigned specific uses depending on prevailing traffic conditions.

Traffic signal engineers have been applying such approaches for many years, with signal timings that adapt to current traffic conditions as measured by sensors and relayed by communication networks to control software and hardware. The ability to make hard measures relate more closely to current operating conditions and thus increase the flexibility and quality of transportation management is one of the strengths of traveler information.

With regard to fiscal persuasion, we see some great possibilities for the application of traveler information systems here also. One important way to influence traveler behavior is to ensure that the traveler has the fullest available set of facts and information about the various routes, modal choices, and timing options available. Information on how to make best use of the transportation network and system has the potential to be a major influence on the way users make use of the system. This information could, of course, include details on how much particular travel options will cost, providing positive reinforcement for any fiscal persuasion that may be attempted.

The power of traveler information also lies in closing the gap between perception and reality. The traveler may have an inaccurate perception of the time, distance, or costs associated with various travel options, which could be corrected through the provision of accurate and current traveler information.

It is obvious that there can be no well-defined boundary between traveler information and traffic management. The hope is that traveler information provided to the traveler has a value, possibly related to a change in travel behavior. Such a change in travel or traveler behavior can be viewed as traffic or transportation-system management.

8.3.3.2   Sharing Sensors

Sensor sharing is one of the areas with the greatest potential for cost sharing, collaboration, and partnership. The transportation network, be it a transit system or a highway system, has a fixed infrastructure or right of way on which the vehicles can operate. This defines the boundaries of the area of deployment for sensor technologies as a series of interconnected, long, thin corridors. In urban areas, these become so dense and closely spaced that they become more like a zone or region of interest. Most of this transportation-network infrastructure, with a few notable exceptions like privately operated toll-road facilities, is operated by public-sector transportation agencies using funding raised from a common source: taxation. These agencies strive to ensure that the quality and quantity of the infrastructure is matched to the demand for use. This involves both the creation of additional infrastructure to meet demand and the operational management of existing infrastructure in the most efficient and cost-effective manner. In recent times, such agencies have deployed or are planning to deploy information and communication technologies to assist in the operational management of the infrastructure. These include one or more of the following:

- *Location sensors:* identifying the vehicle and its location on the transportation infrastructure at regular intervals in the course of the journey;

- *Speed sensors:* returning the instantaneous speed of each vehicle passing the instrumented point in the infrastructure, or the average speed of all vehicles in an instrumented section of the infrastructure, or the average time taken for a vehicle to traverse a defined section of the infrastructure;

- *Video surveillance:* providing the ability to view current conditions on the transportation infrastructure at a remote centralized location, such as a traffic-management center;

- *Human-operated anecdotal data-collection facilities:* courtesy patrol or highway patrol officers, using voice or data radio techniques to relay spoken reports on current conditions back to a central management point.

These sensors make use of a network of communication facilities to transmit their data back to a central, remote monitoring location for processing and action on the resulting information. Mainly used for the purposes of infrastructure or network operational management, this could be perceived as an internal feedback loop, designed to enable the infrastructure managers operating the system to have sufficient information about the process they are trying to manage so that they are empowered to ensure that operations stay within predefined performance measures and operational-management parameters. This is known in the transportation business as traffic management, transit management, transportation-system management, or transportation operations. Most systems like these are focused specifically on the collection and processing of data for uses that are exclusively internal to the operating agency or agencies.

Consider now the operation of a traveler information system. The primary goal is also to collect information about the current status and conditions on the transportation infrastructure and transmit such data back to a centralized processing facility. In the case of traveler information systems, the data is collected as the basis for information destined for the world beyond the realm of the operating agency responsible for the management of the transportation infrastructure. Of course, such data can also be utilized for internal purposes related to the management of the infrastructure due to the high degree of commonality between the needed transportation-management data and the traveler information data. In fact, it is legitimate to consider the development and delivery of traveler information as simply another approach toward transportation infrastructure management since the objective is to positively affect traveler behavior through the ready availability of timely, current, accurate, and reliable traveler information.

Traveler information can also be perceived as a totally separate consumer service business with valuable information products and services being developed and delivered to consumers for a price. This independent business could have its own data collection and information-processing and delivery system, paralleling the operation of similar public-sector transportation-infrastructure-management system components.

The National Architecture suggests that this not be the case and that the data-collection and communication facilities be shared between infrastructure-management and traveler information applications, avoiding duplicate investment and parallel operations. This has the beneficial effect of freeing resources from potentially parallel and duplicative efforts to be applied to make the single, common shared sensor and communication facilities better. The National Architecture indicates this as a course of action because, from a system engineering perspective, this is the only appropriate approach. Remember that in developing a future big picture or system architecture we are attempting to maximize synergy and collaboration and minimize overlap and nonproductive, undesirable competition.

Applying this thought within a more practical application context, it is obvious that the sensor and communication elements of a traveler information system (and perhaps the delivery elements also) should be designed within the context of the overall infrastructure requirements for the region, both in terms of infrastructure management and the delivery of external information about current travel conditions on the infrastructure. The exact nature and format for collaboration between public agencies and between the public and private sectors need to be defined on a case-by-case basis taking account of legacy systems, gaps in coverage, and organizational motivations and capabilities.

8.4   Private-Sector Perspective

The private sector focuses less on attaining transportation policy objectives directly and places much more emphasis on providing a value proposition to the traveler. The development and delivery of traveler information products and services takes place within the context of a much larger and less defined big picture than the public sector. Within the information services and other consumer markets where the traveler information is delivered, there is a wide range of uncoordinated, independent activity. A successful information service provider has to navigate through this in order to establish and operate a business based on the supply of traveler information to the consumer.

8.4.1   The Market for Traveler Information

Most of us are not yet convinced that there is truly a solid market for the commercial provision of traveler information products and services. There are a number of market features that make it a difficult area at the current time.

8.4.1.1   Availability of Data in the Right Quantity and Quality

Unlike weather, stock quotes, and sports scores, there is not yet a comprehensive network of manual or automated data points for the collection of the required traveler information. The cost of deploying current data infrastructure technologies and the sheer effort required to plan, develop, design, install, and operate the data infrastructure means that the public sector is a number of years away from complete coverage. Where coverage exists, the data collected is formatted and structured toward local rather than national needs, making data fusion on the national scale a challenge. This is changing as national and international standardization activities begin to deliver the guidance and standards required to bring data collection to a common base.

8.4.1.2   Willingness to Pay

Many of the business models applied by the private sector to traveler information systems assume that the consumer has a willingness to pay some amount of money in return for access to traveler information services. While we believe that this possibility exists, the current market shows only limited success in this area. It seems to be difficult to follow the typical approach path for new market entries, where a set of initial basic offerings are developed and introduced in order to establish the market and create a revenue stream for developing later, more sophisticated services. Either these initial offerings are not meeting basic acceptance thresholds or standards, or they are not offering the kind of value proposition required to convince consumers to make use of them. Another possible cause may lie in the absence of a fully developed set of information delivery channels. Apart from fixed Internet services, the number and range of comprehensive delivery channel options is still very limited. Handheld devices and in-vehicle information systems are still not mass-market items. The market penetration for such devices is nowhere near that of the ubiquitous cell phone.

8.4.1.3 Public-Sector Competition

The range of business models, both basic and enhanced, that we introduced and described in Chapter 6, provides a strong indication that the development of traveler information systems and business arrangements is at a very early stage. There is a market situation that is reminiscent of many other technology-driven markets: radio, TV, electricity, oil, and rail. Most start with bottom-up activities making strong progress on an individual basis, then reach some kind of critical mass at which point rapid spectacular growth ensues. Getting through the tipping point [3] is the aim of most private-sector traveler information enterprises, but it requires a great deal of resources and focused activity. Unfortunately, the public sector, in many cases, provides publicly funded competition to private information service providers during this sensitive early market stage. This is partly because of the transportation management and travel behavior change possibility we discussed earlier. It could also be because we have simply come to expect free traveler information from our public agencies in return for our tax dollars. In many cases the public sector feels that traveler information is a mission-critical activity that must be under the complete direction and control of the public sector. Here we can see that there are two quite different perspectives on the development and delivery of traveler information to consumers. On the one hand, it can be considered as a commodity that the public should have at public expense because of the community benefits that can be attained through the application of the information. On the other hand, it could be thought of as a commodity that has personal value to the consumer and consequently a candidate for payment for use and delivery by the private sector.

8.4.2   Bundling for Success

Just as there can be no well-defined line between traveler information and traffic management, there is a blurred boundary between traveler information and other information services. Indeed, the blur can be extended to include advertising if you consider an advertisement to be nothing more than some information related to a spending opportunity. From the private perspective there is a less well-defined, but far more extensive architecture or big picture, within which the traveler information business operates. Five major streams of real-time perishable information can be defined as the core of this big picture, in addition to traveler information:

- *Sports information:* scores, news;
- *Stock market information:* prices and news;
- *Weather information:* current and forecast conditions;
- *Advertising information:* products, services.

These information streams already exist and make use of a sophisticated range of delivery channels. It is clear that there is tremendous opportunity to bundle and package traveler information into these same streams. We touched on this in Chapter 6 when we reviewed the range of business models that a private-sector enterprise might consider when participating .n the traveler information market. The use of the portals model would involve the bundling and packaging of information streams designed to target specific market needs. This supports a national versus regional operation.

8.4.3   Managing Potential Synergy with the Public Sector

Although the public-sector perspective on how traveler information systems fit within the bigger picture is quite a bit different from the private view, there is still considerable scope for collaboration and cooperation. As we described in Chapter 6 when we defined a range of business models that can be applied to traveler information systems, the essentials for synergy are as follows:

- Understanding the public-sector drive and motivation;
- Sharing business models and plan;
- Highlighting capabilities and resources.

Through these steps it is possible to minimize the degree of overlap and maximize the probability of success for both public and private participants in a traveler information system initiative.

8.5 Future Possibilities

Looking beyond the current applications of information and communication technologies to traveler information, it is possible to see a new wave of applications that are different from current traveler information applications, but can be related to and integrated with such systems. While not all of the options described may come to pass, we hope to provide additional insight into the wide range of possibilities for applying traveler information system technologies within a much wider context, reinforcing the importance of taking a good look at the wider picture when planning and developing such a system.

8.5.1 Information and Location Systems Convergence

There are a range of location and positioning technologies already in use within the information and communications sphere of influence. These include GPS and the use of cell phones and toll transponders to determine position and location. Using these technologies either independently or in combination, it is possible to locate a person or a vehicle to within about 100m or better, almost anywhere on the globe. Through the convergence of such technologies with those of traveler information, we could see the emergence of a whole new range of position-based information services. In addition to the location technologies acting as sensors for traveler information systems, providing data, such as average vehicle speed and travel time as required for input to the systems, the availability of accurate position data holds out the possibility for new and enhanced services to the end user or consumer. These services also promise to offer new business and revenue opportunities for the private-sector information service provider participants in the traveler information market.

The services in question relate to the customization or tailoring of information according to the location of the user. For example, one of the major issues in ITS these days is the possibility of information overload or driver distraction caused by the communication and delivery of too much information to the driver while he or she is operating a vehicle. A way to address this is to customize the information to the extent that only that absolutely necessary to the driver at that particular stage of the journey is actually delivered, minimizing the flow of information and hence the possibilities for driver distraction. Of course there are other ways to address the issue, including the careful selection of delivery mechanisms. Voice synthesized information could result in less distraction than, say, visual information. In fact, some recent proposals make use of small motors that are integrated into the driver's seat to provide tactile or haptic information delivery. Scientists have discovered that by utilizing a series of such point activators, it is possible to reproduce patterns, such as left, right, and ahead arrows, that can be felt on the back of the driver, thus delivering the information without overloading the visual or aural senses.

However interesting this is, we digress from the main subject at hand—so let us return to it. The customization or tailoring of information to just what is required or desired would also seem to hold out the possibility for enhanced prices and higher revenue. It must be assumed that the driver or consumer of the end product information would be prepared to pay a premium for information that is absolutely relevant to the current context and just in time. This would present the information service provider with one of those elusive opportunities to add value and charge more money. As you can imagine, there is a wide range of enhanced services that might be supported through the convergence of traveler information and position/location technologies. The following sections describe several of these.

8.5.1.1 In-Vehicle Signing

This involves the delivery of in-vehicle information regarding route and destination possibilities within a focused zone around the vehicle.

8.5.1.2 Dynamic Information Resolution

This is a terrible piece of jargon, but we could not think of any other way to describe it. This involves the raising or lowering of the amount of information delivered to the consumer according to the location or to the stage of the journey. For example, if the driver is traversing a long stretch of freeway where there will be no turning or diversion opportunities for several miles then the information stream would be minimal. However, as the driver approaches an urban area and perhaps gets close to the ultimate destination, then the level of information or resolution of the information would increase.

8.5.1.3 Dynamic Trip Assignment

This is a transportation planning or traffic engineering term for a change in the proposed or planned route or mode of travel after the journey has started.

It is assumed that new information is available to the traveler that has caused him or her to change the planned route or the planned mode of travel for the remainder of the journey. The ability to converge traveler information with position information would enable concise mode and route alternatives to be presented to travelers at decision points in the trip. For example, while driving past a park and ride facility on the outskirts of town and heading for the central business district, a single occupancy vehicle driver could be informed of faster, cheaper, and less environmentally damaging public transit alternatives.

8.5.2   Information, Position, and Payment Systems Convergence

It gets really interesting when, in addition to converging information and position technologies, we also converge electronic payment technologies. Electronic payment systems encompass the use of information and communication technologies for toll collection, parking-fee payment, transit fares, and commercial vehicle permitting within the world of transportation. They can also be applied to payment for just about anything in the wider world beyond transportation. This includes paying for burgers, gas, groceries, and a whole lot more.

Imagine, then, the possibilities for even more enhanced traveler information services, incorporating a payment dimension. Taking the dynamic route assignment example once more, we can now charge the driver according to the value of the information delivered. If the traveler information enables the driver to save $10 by taking a different mode or route, then an information service provider could claim a small percentage of the savings as a premium.

Take the case of the BMW driver who is passing the local BMW dealership and is informed that while his vehicle could travel another 1,000 miles before a routine service is due, he will receive a 10% discount if he takes it to the dealer now. This provides the dealer with a new service that can be used to apply demand-dependent pricing strategies. The dealer can review the current and forecast workload for mechanics in the workshop and optimize the pricing structure to make best use of the labor and equipment assets available. This is the sort of value proposition of which new businesses are made. On the consumer side, the driver is provided with the service of being informed of the imminent need for a service, in conjunction with added value information on what opportunities are available and what discounts could be made available for a slight change in plan or behavior.

Consider another example (one of our favorites). My automobile insurance company has decided to take full advantage of the convergence of traveler information, position, and payment systems technologies to offer me a new service known as dynamic insurance rates. Starting on the first of the month, they no longer ask me to pay a fixed premium per time period for car insurance coverage. They have determined that this pricing structure, based on actuarial data approximated by age group, residential district, type of vehicle, and previous driving record, is not accurate enough to give me the best deal or to give them the appropriate level of risk management. Therefore, they have adopted a new technology solution that enables them to locate my vehicle, determine the current risk exposure based on congestion levels, accident statistics, and current driving speed, and then charge me the appropriate insurance premium on a per-mile basis. This suits me since I now have a much higher degree of control over my insurance rates: I can avoid dangerous driving behavior and known congestion points on the network if I choose. The cost of my insurance is more closely related to how I behave, the usage pattern for my vehicle, and the current transportation-network conditions. For example, if I go on vacation for 2 weeks and decide to leave my car in the garage and take a taxi to the airport, then my premiums may drop to close to zero.

On the other hand, I may choose the convenience and flexibility of taking my car to the airport and parking it in a public parking facility there; in which case my insurance rates may be a little higher due to the increased risk. From the insurance company's perspective, the solution offers a much higher resolution approach to the risk versus compensation balance that they manage as part of their business operations. The ability to closely match risk to insurance rates removes the averaging effect of time period average premiums and provides a tailored solution. The ability to reduce overall costs for low-risk drivers and situations and increase overall costs for high-risk drivers and situations provides a much higher degree of flexibility to customize the business model to the market. It also has the desirable effect of creating a motivation for improved driver safety and behavior that can be measured directly with feedback to the driver. Making use of the traveler information components of the solution, the driver can be made aware of current insurance costs and changes in costs on a close to real-time basis. This enables the driver to observe the direct correlation between travel behavior and the cost of insurance.

These are just two examples of how the convergence of traveler information, position, and payment technologies could be utilized to formulate and offer new services and value propositions to the end user. We have offered these visions of the near future to provide a private-sector perspective on how traveler information can and will fit within the future big picture for delivery of information and communication technology supported services. We think we have done something else in the process. These examples, or visions for the future, also serve to reinforce the value of developing and operating traveler information systems in the short term. From both a public and private perspective, they show that if we can work together now and start delivering the basic services to the public in the most efficient and effective manner, then we also purchase future options. We position ourselves for later opportunities to deliver significant value to the traveler and provide the kind of rich information stream that has a high potential to induce travel behavior changes, such as changes in mode, route, and timing of journeys. The very things that public agencies seek to achieve as they pursue their transportation policies and objectives could be attained within the overall framework of a successful traveler information service industry. These future options could have considerable value since they have the true potential to address both public- and private-sector objectives with respect to traveler information systems development and deployment. This leads us to the thought that perhaps we should attempt to evaluate and quantify the magnitude of this value. Maybe we should be making use of techniques from the field of economics to place an agreed value on the future traveler information options that can be purchased by investing in the initial and basic systems today. This added value would, we are certain, justify a much higher level of current investment in traveler information systems, pushing us towards the virtuous circle we discussed in Chapter 6, where improvements in the quality of the offering increase the value of the offering, making more money available to improve the quality of the offering even more.

8.6 Summary

In this chapter we have defined the relationships between traveler information systems and the wider application of information and communication technologies to enable services to be developed and delivered to the consumer. We have addressed this in terms of both public and private perspectives. From the public perspective, we explored the possibilities for synergy, integration, and data- and information-processing sharing and collaboration across public transportation agencies and across different application of ITS. One of the most important aspects of the approach to traveler information system development and deployment relates to the opportunities for sharing and collaboration with other agencies, demanding that such systems not be developed in a vacuum, but with the full knowledge and leverage of the wider context. In fact, cost sharing may be an essential element of any successful approach due to the relatively high levels of investment required to develop and deploy a system capable of delivering credible and valuable traveler information to the user.

In the course of our exploration of the topic, we have come to the realization that information and communication technologies quite naturally lend themselves to collaborative approaches due to the marginal cost of adding capacity or capability. However, this is likely to be counter to the traditional sharp focus on independent initiatives and operations by different agencies. Since this issue is common to many elements of intelligent transportation systems, we made use of the U.S. National Architecture for ITS to explain and illustrate the possible linkages and relationships between traveler information and other systems with a regional or national setting. Again, we want to stress that the U.S. approach to architecture development and application is just one of several approaches that have been defined and applied throughout the world. We made use of the U.S. version as this is the one we are most familiar with and used to make our points with the greatest ease. One of the most important of these points is that traveler information systems have the potential to act as pivot points in wider regional and national ITS deployments. The sensors, communications, information processing, and delivery channels required for effective traveler information can be utilized as a platform or first step to a multifunctional regional or national ITS. To illustrate the possibilities in this respect, we identified some specific opportunities for public-sector collaboration and cooperation from both a functional and a data perspective. Functionally, we identified the close relationship between traveler information and traffic-management applications. From a data perspective, we defined the possibilities for sensor, communications, and information-processing activities.

In our examination of the private-sector perspective with regard to the wider context, we examined the opportunities for sharing information delivery channels with other information types and other information service providers. We also explored the issues associated with cooperation and collaboration with the public sector. Finally, we took a short leap into the near-term future to explore the emerging possibilities for cross-functional convergence between information, position, and payment technologies and applications. We hope that we have imparted to you a clear insight into the important issues and aspects of traveler information systems within the wider information and communication systems context. We believe that developers and implementers of these systems must have such a view or vision if deployments are to be successful in both the shorter term and the longer term.

References

[1]   McQueen, J., and B. McQueen, *ITS Architectures*, Norwood, MA: Artech House, 1999.

[2]   National Architecture for ITS, U.S. Department of Transportation, http://www.itsa.org.

[3]   Gladwell, M., *The Tipping Point: How Little Things Can Make a Big Difference*, Boston, MA: Little Brown Book Publishers, 2000.

9

Summary and Conclusions

9.1 Learning Objectives

After you have read this chapter you should be able to do the following:

- List the essential knowledge, lessons, and experiences described in the book;
- Define a course of subsequent action with respect to planning, deploying, and operating ATIS.

9.2 Introduction

Having almost completed our book and delivered the intended knowledge, information, and practical experience to you, it remains for us to highlight, emphasize, and summarize the most important points and provide some advice for future actions. This chapter provides just such a summary and action list by revisiting the lessons and knowledge provided elsewhere in the book and distilling them into concise points of information and advice. We bring all the essential elements of the book together, providing a review of the knowledge and information we have provided. We also define what we think you should do with this new information, in the form of a series of proposed next steps and further actions.

We begin this chapter with an overview of the essential information provided in previous chapters. This information has been collected under subject headings that relate to the title of each chapter but have been modified to reflect actions required. Then we move on to a brief discussion on some overarching principles and themes that emerge from the combined knowledge and experience captured in the book. Finally, we wrap it all up into a set of concise action items to be used if you wish to apply some of the knowledge and information found in these pages.

9.3   The Essential Points

The following sections represent a structured summary of the essential points and information elements from the various chapters of the book. We have grouped them according to the major themes of each chapter to make it easier for you to refer back to the chapters for more detailed information and discussion on each topic.

9.3.1   What Is ATIS?

With regard to the nature and characteristics of ATIS, we offer the following points of information.

9.3.1.1   There Are Different Levels of Traveler Information Systems Deployed Around the World

Many traveler information systems cannot really be classified as advanced since they have a high reliance on manually derived and anecdotal data. There is a place in the scheme of things for those types of traveler information systems, but we should not fool ourselves that commercially viable content-filling techniques that are more akin to advertising will also satisfy the need for decision-quality traveler information. Anecdotal, subjective information is better than nothing but nowhere close to the information required to support the services depicted in our traveler information vision. Thinking of vision in another way, anecdotal, subjective data can be a bit like using your peripheral vision instead of your primary vision. You see things out of the corner of your eye, you are aware that there is something there, especially if it is moving, you can distinguish the outline shape and have some idea of the color, but you cannot see the detail. It provides context information that alerts you that something is going on that you may be interested in or that may be a threat. It is useful for presence detection but not for detailed evaluation and analysis. When you focus your eyes on something and make of use your primary vision, you can distinguish details and your brain is provided with enough information for you to develop strategies and tactics to react to what you see. Of course this requires more effort on your part and increases the processing load on your brain. Putting this back into the traveler information systems context, anecdotal subjective data can provide the context, the background, and the basic information you need to be aware that something is going on.

Looking at the traveler information scene from a market and business perspective, it is readily apparent that there is an extensive overlap between the business of providing content to support advertising and sponsorship of TV and radio slots and the business of delivering decision-making-quality information to travelers. When you think about it, we are all in the same business of developing and delivering a value proposition to the consumer of the information. In the case of the advertiser, the value proposition revolves around the benefit that can be derived by the consumer from the use or acquisition of the product or service being advertised. From a sponsorship perspective, this becomes a little more indirect since the sponsor is seeking exposure for the value of his or her products and services, through being associated with the delivery of another value or service to the consumer. From the public-sector transportation agency point of view, the value proposition is related to the provision of the information required to enable the user of the transportation system to make best use of it and maximize the value of it. An underlying assumption in the delivery of traveler information to the public is that the availability and accessibility of such information will increase the value of the transportation network by making it easier and more comfortable to use and by increasing the level of benefit derived through use. Both of these public-sector objectives are attained through optimization of the operational performance of the transportation network. When you change your viewpoint to that of the consumer of the information (the traveler), it becomes clear that advertisers, sponsors, and public-sector transportation agencies have one major area of commonality: you, the consumer of the information. This should give us a clue on how best to approach the identification and management of synergy between the various information providers. Through the exploration, definition, and confirmation of the precise nature of the value we are all trying to deliver to the consumer, we should be able to develop the basis for coordinated, collaborative win-win approaches. This should take account of the relative strengths and weaknesses of the various public- and private-sector partners.

9.3.1.2   A Higher Level of Automation and the Delivery of Decision-Quality Information to the Traveler Is Vital

Our ultimate goal must be the development and deployment of ATIS that have a higher level of automation and deliver higher quality objective information to the traveler. This is the only way to address the common ground between the public and private sectors. Higher-quality data can support the public-sector need to provide decision-making information that will influence and inform the traveler while at the same time providing the private sector with the raw materials on which to base successful, sustainable businesses.

9.3.1.3   Predictive Traveler Information Needs To Be Part of the Future

Another assumption we made when we painted our picture of the rosy future with full access to traveler information services was the availability of high-value information, such as route and mode alternatives and predictive traveler information. Information that is even a few minutes old is of much less value than an accurate forecast of travel conditions in the next few minutes. Just like weather information services, while current conditions are of great interest, the focus of interest is on what is likely to happen next, rather than what has just happened. Continuing the weather service analogy, it is obvious that the accuracy of the prediction is directly related to our detailed understanding of the fundamental mechanisms that drive and control conditions and the availability of enough decision-quality data for us to measure the parameters within such mechanisms. Considering the mechanisms first, we are very fortunate in the transportation community to have a long and successful track record in the study, identification, and understanding of the factors and mechanisms that drive travel demand and dictate travel conditions on the transportation network. We understand the fundamental relationships between economic activity, population demographics, and travel demand. We know, in great detail, the relationship between demand, capacity, travel speed, and congestion. We have developed sophisticated tools and techniques that enable us to study such effects, replicate measured travel conditions, and make future predictions. When you take a look around the world, there are almost 100 mathematical simulation models that have been developed as tools for the study and evaluation of transportation networks and the evaluation of the likely impacts of future transportation facilities and plans. Unfortunately, when you dig a little deeper and look at the factors that constrain the accuracy and usefulness of such models, a familiar theme emerges: garbage in, garbage out. The lack of sufficient decision-quality data is the major constraint in the accuracy of the predictions that emanate from such mathematical simulation models. The bottom line on all of this becomes relatively simple. If we build the data-collection infrastructure we need to support traveler information and transportation-network management, we will also enable the provision of more sophisticated and hopefully higher-value traveler information services, such as travel time forecasts and transportation-network conditions predictions. We will also equip ourselves to conduct more effective trans- portation and land use planning through the ability to make more accurate assessments of the effects of development and urban land use on the surrounding transportation network. The availability of a reliable stream of decision-quality data will also equip us to discover more about the fun- damental mechanisms that are present in our transportation network and enable us to devise more effective approaches to the management of these mechanisms. In simple terms, more and better data leads to more effective and efficient management of the transportation network, as well as better use of the network on the part of the users.

9.3.2 Effective Utilization of Traveler Information System Technologies

To help you to select and deploy ATIS technologies as effectively as possible, we made the following major points.

9.3.2.1 There Are Two Major Groups of Traveler Information Data Sensors

Two major groups of sensors are available for utilization within ATIS: point sensors and area sensors. Point sensors, as the name implies, collect data at a single specific location on the transportation network; while area sensors provide data about a region or area of the transportation network. Both types of sensors have a vital role to play in the collection of data for input to ATIS. Point sensors can provide detail on the types of vehicles and the instantaneous speeds of vehicles on the networks, while area sensors can deliver journey time data that can provide an overview of network status and the basis for predicting future network conditions.

9.3.2.2 Be Aware of the Potential for the Emergence of a Disruptive Technology

Technologies for point and area sensors, telecommunications, information processing, and information delivery are dynamic, with new ones emerging at an increasingly rapid pace. This makes it very likely that a disruptive technology could emerge and change the nature of the market. It is impossible to predict the nature and timing of such technologies, but we would expect breakthroughs in area-wide sensing technologies and techniques, such as the use of cellular telephone position information to determine travel times of vehicles and the collection of data from vehicles in so-called super probe applications.

9.3.2.3   Interactive Voice Response Technologies and the U.S. 511 Initiative

In the near term, the widespread application of interactive voice response technologies for flexible access to traveler information from land lines, and static and mobile Internet will be driven by a number of government initiatives now underway. A good example of this is the 511 initiative in the United States. On July 21, 2000, in response to a petition from the U.S. Department of Transportation, the Federal Communications Commission assigned the number 511 as a nationwide telephone number for traveler information. This means that once the infrastructure is in place, you will be able to dial 511 anywhere in the United States and be given access to a range of traveler information services relevant to the local region in which you made the call.

The 511 initiative is worth another look from a different perspective. One of the challenges facing both the public and private sectors as they develop and deploy traveler information services is the identification and application of suitable techniques to raise awareness of the availability of the services and the benefits to be gained through their use. Product and service developers in other markets make extensive use of marketing, advertising, and awareness campaigns to raise such awareness, grow consumer interest in acquisition and use, instill a desire in the consumers for acquisition and use, and finally move the consumer to some type of buying or acquisition action. The 511 activities in the United States, in addition to spurring the technical institutional and organizational actions required to establish the services, has also developed a major awareness and interest focus on the subject of traveler information. The 511 initiative is beginning to take on the attributes of a brand that can be used to instantly communicate certain values and attributes to the consumer. It is acting as a focal point for the various activities designed to develop and deliver the desired services, while providing an excellent story line about traveler information and the national need for it.

9.3.2.4   Potential for the Application and Use of Standards

Through the use of the traveler information supply-chain and coherent business-planning approaches, it is possible to develop a conceptual framework or architecture for the proposed traveler information system infrastructure and delivery mechanisms. This big-picture view enables the identification of synergy with other systems and applications, but also provides the basis for the identification of appropriate standards to be applied at the various steps and links in the chain. Here again, the traveler information supply-chain concept can be utilized as a tool to consider standards application in a systematic, structured, and thorough manner. Taking each of the seven steps in turn, a review of available standards can be conducted and appropriate standards can be identified and incorporated into plans, designs, and business concepts.

9.3.2.5 A Dynamic Technology-Development Environment

There is a wide range of information and communication technologies available to support the deployment of ATIS. These are developing at a rapid pace, and they create a dynamic development environment.

In Chapter 3 we reviewed the technology options and possibilities related to traveler information systems. While we are fairly certain that we did not cover all possible options and that by the time this book is published there will be new options and technologies, that is not the point. The main message we want to deliver is that this is a dynamic field where most of the technologies are being sourced from efforts and markets that have nothing to do with transportation. To make sure that you do not overlook a technical approach or technology, it is important to conduct a thorough review of technologies currently available and those that may be available in the near future. This should be complemented by an investigation of the lessons and experiences derived from other recent deployments, ensuring that the full benefit of this experience is incorporated in the chosen approaches. It is likely that the investment required in data-collection sensors, telecommunications networks, and information-processing and delivery facilities will be substantial, making an initial investment in technology review and analysis a very worthwhile effort that will deliver considerable value.

9.3.3 Exploring and Managing Public Objectives, Private Enterprise

With respect to the identification, exploration, confirmation, agreement, and management of traveler information system objectives for both the public and private sectors, we offer the following essential points.

9.3.3.1 There Is a Range of Public-Sector Organization Roles and Types Present in Any Region and Organization with Respect to Traveler Information Systems

A wide range of public-sector agency types and job roles and responsibilities are present in any region. We have provided an initial assessment of the types that might typically be present as good base material for subsequent development of effective business and technology application plans. In Chapter 5 we reviewed a range of possible public agency types and the types of roles and responsibilities that might be found within such organizations. These can be used as a starting point or reference list for the development of your own region-specific list.

9.3.3.2 Developing a List of Public Agency Transportation Objectives

Considering your agency and the other regional public agencies, it is possible to identify, define, and agree upon a set of public transportation policy objectives that are suitable to be addressed by the application of ATIS. This is a crucial step in the development of business plans for traveler information services and product development as this information is used to guide and drive subsequent steps in the planning, deployment, and operations process.

9.3.3.3 Defining the Range of Private-Sector Organization Roles and Types with Respect to Traveler Information Systems

In the same way in which pubic-sector organization types and organizational roles and responsibilities have been defined, it is necessary to also define the private-sector enterprises and roles and responsibilities that may impact the successful planning, deployment, and operation of ATISs within a region.

9.3.3.4 Defining a Set of Private Enterprise Traveler Information System Objectives

While most private-sector enterprises will have defined their internal goals and objectives, these may not have been communicated to the relevant public-sector organizations. It is important that the public sector reach out to the private sector and understand these objectives and motivations, contrasting and comparing them against public objectives. Only then can the partnering process begin.

9.3.3.5 Distinguishing Real Objectives from Metaobjectives

Standard system analysis and requirement definition techniques can be applied to ensure that the initially stated needs, issues, problems, and objectives are indeed true core requirements and not some sort of higher-level metaobjectives. It is vital that the effort is made to reveal and agree upon the true requirements that directly relate to both public transportation policy objectives and private-sector business objectives. Settling on metaobjectives will lead to a suboptimal plan for traveler information systems and could jeopardize the chances of success.

9.3.3.6 Defining at an Outline Level the Potential Areas for Public-Private Collaboration in ATIS Activities

In order to facilitate the establishment of an effective dialog between public- and private-sector entities, it is very useful to have at least an outline concept of the areas for potential collaboration. A high-level definition of potential collaboration areas helps enormously in the identification and assessment of potential partners. Once you have an idea of the areas where you wish to seek collaboration, you can identify suitable partners in the market, analyze and evaluate their technical, organizational, and business capabilities, and determine their overall suitability. As we stated in Chapter 6 when we introduced the basic traveler information system business models, these can be utilized as a tool for defining and exploring potential areas for collaboration and partnership frameworks. We would also like to offer the public sector some additional advice in the form of a simple process we have used to good effect on previous work. It is an abstraction of the proposed public-sector business-planning process for ATIS described in Chapter 7 and is illustrated in Figure 9.1.

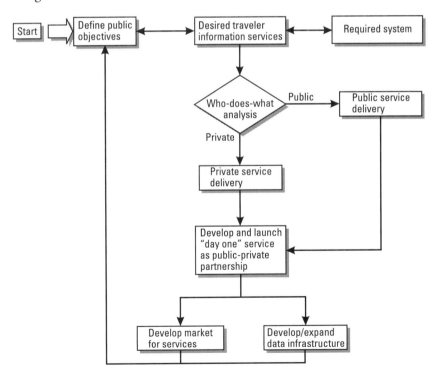

Figure 9.1 The who-does-what diagram.

The process starts with the definition and agreement of public-sector objectives associated with ATIS, leading to a determination of the services and the system required. These first three steps are likely to be iterative in nature as changes in the perception of need caused by awareness of systems capability will, in turn, affect system configuration. The who-does-what analysis is then conducted, which involves a review of what public- and private-sector agencies can provide by way of capabilities and resources. Once this has been carried out and a detailed business plan as described in Chapter 7 has been assembled, it is possible to launch day-one services supported by an initial public-private partnership. Parallel development of the infrastructure and the market for services leads to evolution of the system and the operating context over time; this leads to the need for reevaluating the technical and business structure of the ATIS by repeating the process.

In the who-does-what analysis decision box, we recommend the use of what we refer to as the wobbly triangle diagram, as illustrated in Figure 9.2 This represents the total cost of the subject ATIS as the area of a triangle. The base of the triangle along the horizontal axis represents the private-sector contribution to the total cost of the system, while the vertical axis represents the public-sector contribution. The use of the triangle enables the total cost

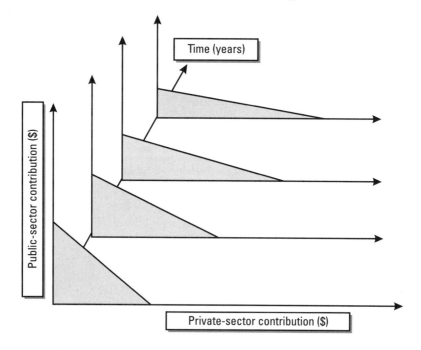

Figure 9.2 The wobbly triangle.

and the expected contributions of each sector to be graphically depicted in terms of the day-one situation and planned future year arrangements, as illustrated by the other triangles shown behind the first one along the time axis. This takes account of the potential for partnership arrangements to be dynamic over time.

9.3.3.7 Defining and Exploring the Possibilities for Synergy Between Private-Sector Businesses

As we discuss the possibilities for collaboration between the public and private sectors, or intersector collaboration, we should also be aware of the possibilities for intrasector collaboration between different public organizations and between different private enterprises. In Chapter 6 we identified and explored some of the possibilities for public agency collaboration in traveler information within the context of big picture planning for ITS at regional and national levels. We believe that the most effective approach to the application of information and communication technologies to transportation-network development, operations, and management involves the initial development of a high-level conceptual framework or architecture that defines the needs and solutions in a coherent way that supports the identification of major technical components and relative roles and responsibilities. The development of a regional or national system architecture for ITS is an excellent first step in the identification and definition of the possible intersector relationships and joint activities with respect to traveler information systems.

Looking from the private-sector perspective, the development of services and products based on information and communication technologies holds out more opportunity for alliances, networking, and collaboration than most other business sectors. As witnessed in the development of the huge number of Internet businesses engaged in the delivery of information and the retailing of products using the Internet as the sales medium, there are almost unlimited possibilities for synergy and collaboration. The ability to link your Web site to that of another enterprise at low cost and with very low technical difficulty supports the culture of collaboration and sharing. We have seen some incredible examples of the establishment and successful implementation of symbiotic relationships between enterprises that either sell to the same customers or market segments or have identified ways in which they can market and sell their offering while marketing and selling the offerings from others. In many cases, the identification and selection of the opportunities to work and invest collaboratively comes from a solid investigation and analysis of the value chain or supply chain in which each enterprise is participating. This enables the enterprise to define the business it is in, the activities it needs to support, and the relationships it needs to develop with the other participants in the chain. For example, a private-sector enterprise that is in the business of selling buses to transit agencies may identify an opportunity to collaborate on business development and sales activities with another enterprise trying to sell on-board electronics to the same customers. While the best companies have tended to take more of an independent approach, perhaps by developing their own organizations to incorporate the additional products and services, the trend today seems to be toward specialization and focus on what your enterprise does best, using networking, alliances, relationships, and partnerships to provide the completeness or comprehensive offering required to fully meet customer needs, while maximizing the efficiency and effectiveness of business operations.

9.3.3.8 Making Sure That You Really Know What You Are Getting into When You Launch a Traveler Information Service

In our experience, most public-sector transportation agencies have tremendous resources and capabilities with respect to the planning, deployment, and operation of transportation infrastructure. In most cases, however, this involves very little in the way of an ongoing dialog with the users of the transportation facility. Once the new infrastructure is in place, whether it is a new road or a new transit facility, there is typically a drastic reduction in the two-way communication between the operators and the user. There are, of course, exceptions to this rule in the form of enlightened public transit agencies and toll road operators. On the whole, the average public agency deploying the typical asphalt, concrete, or steel transportation infrastructure has little need to open and maintain a long-term dialog with the facilities users. Information and communication technology applications tend to be very different in this respect. The delivery of information on how best to use the transportation network and the options available to the user requires the establishment and ongoing management of a two-way communication between the system operator and the user. This operational dialog requires some very specific skills and resources if it is to be supported efficiently and effectively. These are more akin to those found in service operations and consumer-oriented companies, than in public-sector transportation agencies.

9.3.4 Application and Use of the Traveler Information Supply-Chain Concept

When it comes to the application and use of the seven-step advanced traveler information supply chain defined in Chapter 6, we make the following primary observations.

9.3.4.1 The Traveler Information Supply-Chain Concept

To help in the planning, development, and subsequent deployment of traveler information systems, we can consider the entire span of events as a business process or supply chain consisting of seven primary steps that can be used to model business opportunities and explore relative roles and responsibilities for both public- and private-sector entities. The seven steps, introduced in Chapter 5 and described in detail in Chapter 6, are as follows:

1. Building a data-collection infrastructure;
2. Collecting data;
3. Data fusion;
4. Information processing;
5. Adding value;
6. Marketing;
7. Delivery.

Each of these steps in the process supports cumulative activities and investments that support the development and delivery of the required traveler information products and services at the end of the chain. The chain can be considered as a model for the definition and exploration of alternative activities and investments, the study and understanding of the effects of such actions, and the mechanisms for cooperation and partnering that exist. Thinking of the process as a sequence of focused actions enables detailed analysis on each part of the overall chain, and then reassembly of the elements into the overall whole.

At each point in the value chain, the appropriate roles for public and private participants can de defined and explored. Splitting the overall chain into smaller, more manageable units supports a greater focus on each element and the assignment of work activities to a team of people, enabling specialist resources to be effectively applied and detailed analysis work to be accomplished. The relationship between the various steps can also be examined in terms of the data or value that will flow between each step, enabling the revenue, value, and data flow from end-to-end across the process to be identified, defined, and examined. This ability to focus on the details while maintaining a clear view of the bigger picture and overall context is the primary value delivered through the use of the value chain model. Note that relevant roles depend on the specific context and the starting point from which the development activity begins.

When you take the wider view and consider the entire traveler information tion supply chain, it is very easy to identify opportunities for public- and private-sector collaboration. Common ground can be identified in almost every one of the steps in the chain, particularly with respect to data fusion, information processing, and the establishment, deployment, and operation of information delivery channels and mechanisms. It is apparent that the public sector requires the assistance of the private sector, especially if all effective information delivery channels are to be fully utilized. It is also apparent that the private sector will struggle to develop financially feasible and sustainable businesses unless they have access to the right quality and quantity of publicly derived data regarding the status and condition of the transportation network.

9.3.4.2   Business Planning and the Development of Business Models for Traveler Information

In order to ensure that the possibilities for cooperative action and effective collaboration are both understood and exploited, activities right at the beginning of the development process need to be integrated and viewed in parallel. The business planning associated with public- and private-sector traveler information systems initiatives provides wonderful materials and information that could form the basis for effective dialog and further collaborative planning efforts in the traveler information systems domain. Through the adoption of a structured approach to such plans, both sectors would facilitate and enable the sharing of information and the development of a common understanding of requirements and motivations. While we have suggested an outline business-planning process for each sector to try to encourage such interaction, the critical factor is that the planning process for each sector should incorporate outreach and partnership activities from the ground up.

We also advocate the adoption of user-needs-driven approaches to the planning and development of traveler information systems. This ensures that clear definition and agreement of the needs, issues, problems, and objectives to be addressed is achieved as the basis for success. From a public-sector perspective, this avoids the adoption of false or metaobjectives, while providing a clear view of the relationship between actions and investment and initial objectives. This should also highlight the fact that the public sector, in order to be effective in traveler information delivery, must understand and manage a new paradigm in which they invest and participate in new business domains and market areas beyond their span of control. Influence and advocacy skills become paramount, and the public-sector participants must master partnership and alliance-building techniques beyond direct regulation and direction.

From the private-sector perspective, this creates a sharp focus on what the end user or consumer is perceived to need and want and a clear picture of the business model and operation proposed for the enterprise.

9.3.4.3 Business Models

In Chapter 6 we explored the concept of business models. While the concept is not new, we believe that the widespread application of the business model concept to the planning and development of an ATIS brings a considerable amount of value. Defining the way in which an enterprise will do business and how it interacts with other entities in the traveler information supply chain and the capabilities and resources that make it unique and provide unique selling points brings an understanding of organizational objectives and relative positioning with respect to other participants that is vital to success.

An understanding of the principles behind business models also allows the traveler information community to look beyond its own boundaries for ideas and proven approaches to the development and delivery of advanced information services. External sources, such as the Internet business area, utilities, and fast food can all provide a depth of information and ideas for concepts that can be applied to traveler information.

Incorporating the definition and refinement of the appropriate business model within the overall structure of a formal business-planning approach for both public and private sectors raises the value further still. The development and confirmation of requirements and objectives, coupled with the identification of activities to be supported and relative positioning in the value chain, supplies the framework for collaborative approaches to the delivery of effective traveler information. This also provides the public-sector participants with the opportunity to define approaches and techniques to address the wider domain of information services and consumer products in which they now find themselves participating. Going beyond the traditional procurement of goods and services, it is necessary to identify effective strategies and tactics to communicate, influence, and advocate in the wider domain.

9.3.4.4 The Range of Downstream or Delivery Business Models Available

We cannot overemphasize the importance of taking a broad view of technical, organizational, and business aspects of ATIS. The adoption of a narrow approach based solely on the experiences and capabilities of the transportation profession will most likely lead to the reinvention of the wheel and a suboptimal solution. This is particularly true of business models. The Internet and e-commerce business sectors have developed and applied a range of innovative business models that could have direct relevance to traveler information. We introduced and discussed some of these in Chapter 6, but we encourage you to scan for other examples and to be prepared to forage for business model variants and practical experiences in the application of such models before settling on the best models for your context and starting point. The whole information services market is a rich source of inspiration and experience in this respect, so always look within and beyond the transportation realm when studying the business model options.

9.3.4.5 Considering Tools, Technologies, and Techniques from Other Business Areas and Disciplines

As we endeavor to explore the value propositions that represent the common fertile ground for partnerships and collaboration, it is useful to keep in mind that techniques and methodologies for exploring, defining, agreeing upon, and quantifying the individual values already exist. One of the most significant features of ATIS (and ITS for that matter) is the direct relevance of tools, technologies, and techniques from other business areas and disciplines. We are constantly amazed by the potential for synergy, cross-fertilization, adaptation, and technology transfer that exists in the application of information and communication technologies to transportation. In this case, a technique known as decision analysis, derived from the world of social sciences, can be usefully applied as a structured approach to the identification and agreement of the value propositions associated with ATIS [1]. This technique supports a formal quantification and analysis of the value that people and organizations place on things, enabling a rational comparison of alternatives and providing a good set of support data for decision making with regard to negotiation and partnership.

The dialog and information exchange required to support the effective establishment and subsequent management and stewardship of a partnership for traveler information creation and delivery goes beyond the definition and agreement of the value propositions involved. In order to maximize the probability of success, the activities involved in the establishment of such dialogs and the sharing of needs, issues, problems, and objectives should be couched within the support framework of an overall business-planning approach to ATISs. While there is a need for both public and private partners to conduct some initial independent analysis and consensus work towards preliminary business plans, at the appropriate point in the development cycle, cooperative planning should begin. We should try to embrace both businesses by developing top-down approaches to ATIS planning, development, operations, and management.

9.3.4.6 Determining What the Users Want from a Market Perspective

This new domain in which traveler information service and product providers find themselves participating has an important consumer-oriented dimension. If you are to have the maximum confidence that your services and products will be acquired, utilized, and deliver value to your target customers, it is important to have a reasonably good idea of what the customer does and does not want. We consider this to be an element of the marketing step in the traveler information supply chain, although it is obvious that some activities would have to be carried out at an earlier stage in the chain as the services are being defined and the products to be developed are identified. To accommodate this, we advocate the use of a formal business-planning process as described in Chapter 6, where an initial analysis and evaluation of the market for goods and services is conducted as a precursor to the business activities that lead to product and service development. The paradox in these activities lies in the fact that the existence of a good traveler information system would make the job of collecting the required market data much easier. The establishment and management of an ongoing two-way dialog with traveler information product and service consumers is fundamental to the successful operation of such systems. The dialog would also serve as the perfect vehicle for identifying and understanding customer needs and consumer reaction to various product and service types. Like many other ITS applications, the system that supports the delivery of the service can also support feedback data that describes how users are actually using the system and how they respond to different service and product options.

9.3.4.7 With the Right Model, the Private Sector Can Make a Business from Supplying Traveler Information Systems

One of the things that our exploration of traveler information system business models has taught us is that the identification and development of the most appropriate model for your specific circumstances has a huge effect on your chances of success. A systematic analysis and evaluation of the core business models and the range of procurement and partnering mechanisms variants we described in Chapter 6 can yield considerable value in this respect. The core models can be used at the initial planning stages to approximately identify and define relative roles and responsibilities, while providing the clear path from requirements to proposed actions and investments. Building on the initial output from the core model analysis, detail regarding the exact nature of the relationships with other agencies and partners and the structure of the formal agreements can be developed with reference to the variants.

While there has been limited private-sector success in the traveler information services development and delivery business in the United States, examples in Europe and Japan indicate that success is possible with the right approach. In Britain, Trafficmaster® has blazed a highly successful trail as a private enterprise focused on the development and delivery of in-vehicle traveler information services. As described in Chapter 2, based on an initial agreement with the British government for the right to install and operate a traveler information data-collection infrastructure on the U.K. motorway (freeway) network, in return for the supply of data back to the government, they have developed a thriving enterprise that provides traveler information services via a range of delivery channels. Geographical coverage has also evolved to cover not just the entire national motorway network, but also significant stretches of major the arterial network (or trunk roads as they are known as in the United Kingdom). This particular example of ATIS development and deployment with solely private-sector financing and direction highlights the value of a solid initial investment in the data-collection infrastructure required to support viable current and future services. It also serves as an example of a successful private-sector approach to the development and sustainable operation of a business based on the delivery of traveler information services.

In Japan, the central government has chosen to take a contrasting approach, featuring the extensive use of public investment and activity to drive the development and deployment of a national traveler information system. The public dominance model has been proven in the Japanese context (VICS).

9.3.4.8 The Planning Payback and the Thought Investment

We cannot overemphasize the value of well-structured and appropriately resourced planning for ATIS. It is not a stand-alone application, but rather is inextricably linked to and overlaps with public-sector traffic-management and private-sector consumer-information products and services. While you could take the view that you cannot possibly hope to structure the perceived chaos that ensues, our experience shows the immense value of having a good plan, relating to clear, measurable objectives. The empowerment that this provides is truly amazing. While the barriers to progress may seem formidable, you have at least identified and defined them, you know what you want to achieve and have at least a preliminary outline of how you plan to achieve it. This positions you for effective dialog and partnership formation, while readying you for any opportunity that may present itself. You have your ducks in a row and can clearly and cogently communicate what you want and how you plan to get it. While you may change your mind, once you understand what others in the public and private sectors are aiming for, you have an invaluable frame of reference. Every dollar spent in the planning and thought phase of your initiative is probably worth at least $1,000 spent later.

Traveler information system development and delivery is not the same-old, same-old for public-sector transportation agencies. Consequently, it is vital to build an effective awareness of the new context in which such systems are developed and deployed. This is not your father's transportation scene in which you hold all the bargaining chips and have all the control and regulation required to make your wishes come true by direction. This is a new, scary context in which you can point in the right general direction, but must expect to be swayed and influenced by external factors and forces beyond your control. The reverse is also true, you will have the opportunity to influence, but not regulate or direct, other significant players, but you have to be good at the game. What is the name of the game? Advocacy, marketing, lobbying, selling, influencing, cajoling, canvassing—call it what you will. The essence of success lies in being able to define what you want, communicate it to others, and demonstrate the mutual value. Do you know enough about your new potential partners to do this effectively?

9.3.4.9 Objective and Open Evaluation of the Value of Traveler Information

In the course of studying the issues associated with the delivery of advanced traveler information services and applying our experience in the field, we have identified a recurring theme. From both public and private perspectives, success in traveler information depends on a number of factors related to the investment of the appropriate level of funding and the identification of the most appropriate business model or mode of operation for public organizations and private enterprises. It strikes us that a clear, unambiguous, objective, and widely agreed upon definition of the value of traveler information is central to successful choices. From a public perspective, the availability of a robust value of traveler information provides the solid economic justification for public investment in the required infrastructure. From a private-sector perspective, such a value represents the financial basis for setting research and product development spending levels and the eventual target or market price for products and services. From both perspectives, the value of the traveler information derived from the system is a primary driver in the selection of business models and overall approaches to the development and delivery of the desired services.

As discussed in Chapter 6, some initial work on the subject leads us to believe that there may be a range of values of traveler information, rather than one single value. At the highest level there may be two sets of value that can de defined: intrinsic and perceived. The intrinsic value of traveler information would be determined on the basis of the costs incurred in developing the required data-collection and information-processing facilities. The perceived value would be determined through the conduct of a benefits analysis, which would determine the perceived value of time savings, pollution reduction, and other performance measures reflecting the value of the information to the traveler and the value of the information to the public agencies involved, by way of degree of satisfaction of transportation objectives. This latter value is, of course, an indirect value to the traveler since the transportation policies and objectives will have been derived from community benefit objectives in the first place. This gives us a single intrinsic value of traveler information and a set of two perceived values as follows:

- *Core value:* the intrinsic value of the traveler information directly related to the cost of collecting and processing the data and delivering the information services and products;
- *Private value:* the perceived value of the traveler information related to the value of the information from the consumers' point of view;
- *Public value:* the perceived value of the traveler information related to the degree of transportation objective satisfaction from the public-sector agency perspective.

While there would be a considerable amount of work involved in developing values for all three, this would provide an excellent basis for the justification of both public- and private-sector investment and activity in ATIS development, deployment, and operations.

9.3.4.10 A National Data-Collection Infrastructure Business Model

An intriguing approach to the definition of a business model for public- and private-sector collaboration in traveler information services delivery is one based on the experience gained in the development and deployment of the Internet. This model incorporates a two-stage approach in which substantial initial investment on the part of the public sector is justified by clearly defined goals and objectives. Such objectives are specific to the public sector and are the sole justification and rationale for public investment of taxpayers' money. The resulting infrastructure delivers public benefits and impacts of sufficient level and magnitude to stand alone as a publicly funded and operated initiative or system. At a later stage in the development of the infrastructure, the commercial opportunities that have been enabled by the initial public investment are realized through the transfer of significant proportions of development and operations to the private sector. In return for the initial public investment that provided the critical mass required in terms of infrastructure and users, the private sector improves the effectiveness, efficiency, and coverage of the infrastructure, delivering additional services and supporting the original public ones required to deliver public good.

Our thinking is that there is a parallel model or approach that could be applied to the world of traveler information. Starting with the initial public investment, this would be in the deployment of sensor and telecommunication facilities required to deliver a nationwide, comprehensive set of transportation data. While this would be initially focused on the delivery of the data set required to support the derivation of performance parameters associated with transportation network-management applications (such as freeway, urban surface street, arterial, freight, and transit-fleet management), it would be designed to incorporate inherent flexibility for future use as the source for traveler information data.

Once the infrastructure is up and running and providing a satisfactory level of data in support of transportation network-management functions, the second stage would be initiated. The fused and processed data from the sensor and telecommunications infrastructure would be made available to the private sector on an open access basis, fostering a host of traveler information services.

9.3.5 Consideration of ATIS Within the Wider ITS Context

When it comes to the consideration of the relative position of ATIS with respect to the wider ITS and transportation context, we offer the following major points.

9.3.5.1 The Absolute Importance of Decision-Quality Data

Not to put too fine a point on it, but the lack of quality, objective data places traveler information systems around about the time that medical practitioners were using leeches as a cure for everything. We are in the dark about the actual performance of our transportation networks and can only make educated guesses about what is going on and how best to manage and react to it.

While this is bad news for public and private organizations interested in the delivery of traveler information services to the consumer, it should also be of considerable concern to public agencies responsible for effective management and operation of the transportation network. If we do not have the data stream required to support effective delivery of traveler information, then we most likely do not have the raw data required to drive and support our transportation network-management efforts. Both applications, traveler information and traffic/transit management, can thrive on a common base of high-quality, comprehensive data regarding the current performance and capacity of the transportation network. There is a very close relationship between transportation-network management and traveler information activities, to the point where it is often hard to differentiate services designed for transportation-network management from those designed for traveler information delivery. One of the central tenets in our overall exposition on this subject is that public agencies should justify public investment in the development of traveler information services on the basis of the transportation policy objectives that can be addressed and satisfied. Many of those objectives will relate to improvement in the effectiveness and efficiency of transportation networks and facilities and be held in common with parallel activities in transportation-network management.

So, what is the bottom line? We advocate the analysis and evaluation of opportunities and possibilities for the public and private sectors to work together in a collaborative, cooperative manner in the development and deployment of advanced traveler information services at local, regional, and national levels. If, however, it proves to be the case that there is no viable means to engage private-sector funding, drive, and talent, then the public sector still has the burden of making something happen. The case for public-sector investment and activity in the delivery of traveler information to its constituents is clear and has been underwritten by precedents that stretch back as far as the installation of the very first road sign. While it is vital to explore the possibilities for public and private cooperation in order to avoid duplicate expenditure and the reinvention of the wheel, there is just too much by way of public benefit to be gained to make a do-nothing approach an acceptable option. We believe that in the not-too-distant future, the lack of an effective traveler information system within a region will be viewed in the same light as a lack of road signs, road markings, bus stops, and station signs would be today.

9.3.5.2 Advertising Traveler Information Services

It is also necessary to define the methods and techniques you will use to make the customer aware of your products and services and understand the benefits and value of acquiring and utilizing them. This is known as advertising. In most cases, public-sector agencies do not have the capabilities or experience that relevant private-sector enterprises have in this area. This could be a significant area for public-private collaboration, even if the traveler information services are to be provided at no cost, since the user still needs to be informed of the arrangements to access the information and the value of making use of it.

9.4 Recommendations for Future Actions

Now that you have read this book, what should you do next? In an ideal world, we would have given you enough information and ideas to inspire you to pick up the lessons and knowledge from the book and apply them in real life. How should you make a start if you are inspired to action? In Chapter 1 we presented you with our vision of transportation's future incorporating ATIS. This illustrated the application of information and communication technologies towards the provision of a range of integrated traveler information services. In order to progress to this desired future state from the current limited availability of ATIS, there are a number of actions needed on the part of both the public and private sectors. We thought that a good way to conclude the book would be to develop a final section that takes all the knowledge and information we have provided in the book and distill it into a concise set of recommendations for next steps or further actions. When we developed this set of actions, we had in mind local public-sector implementers and national policy makers, as well as private-sector information service providers and information delivery device developers, since the most appropriate roles and responsibilities vary according to who you are.

Table 9.1 summarizes our proposed set of five next steps. To indicate which steps are appropriate to each public- and private-sector partner, we have indicated in the table whether the specific organization types should lead the effort or be a collaborator in the effort. Each step is described in a bit more detail in the following sections.

9.4.1 Initial Business Planning

Conducted on an independent basis by public- and private-sector participants, this part of the planning phase is focused on identifying, confirming, and agreeing upon internal needs, issues, problems, and objectives to be addressed by the proposed ATIS initiative development or deployment. This

Table 9.1
Next Steps

Proposed Actions	Public		Private	
	Local Implementers	National Policy Makers	Information Service Providers	Device Developers
Initial business planning	Lead	—	Lead	Lead
Market study	Lead	—	Collaborate	Collaborate
Reference proposed activities to the regional ITS and transportation context	Lead	—	—	—
Detailed business planning and modeling	Collaborate	Collaborate	Collaborate	Collaborate
Traveler information supply-chain management	Collaborate	—	Collaborate	—

provides the raw material to initially justify further actions and support the dialog required for subsequent partnering and detailed collaborative business planning. A critical element in initial business planning is the identification, exploration, and conformation of needs, issues, problems, and objectives through the application of formal requirements analysis techniques. This helps to maximize the chances that you will get what you want by identifying, confirming, and agreeing upon a clear statement that defines what you want. This enables you to ask for what you want in an unambiguous way. This includes getting to the real roots of what you want, drilling down to the fundamental transportation policy objectives that you seek to address, going beyond the superficial. This also enables you to place a value, however approximate, on getting what you want. This supports the simple sequence of ready, aim, fire. We have a client who regularly practices the sequence, ready, fire, aim. It is our job, apparently, to move the target, to ensure that he always hits the bull's eye. Such is the lot of a consultant! Seriously though, you ignore this simple sequence at your peril. We have witnessed many initiatives where ready, fire, aim has been initially adopted as the sequence and the result has been a longer and costlier sequence of ready, fire, aim, fire, aim, fire, aim, fire! Need we say more? ATIS have so much potential and are too important to be treated in a spontaneous, ad hoc manner. Also, ATIS are so closely related to transportation-management applications, such as traffic management and transit-fleet management as well as passenger information systems that we cannot afford the bottom-up technology-islands approach.

9.4.2 Market Study

This would also be a parallel activity on the part of both the public and private sectors. From a public point of view, the objective of the market study is to identify the private-sector partners that may be present and available as potential partners. The initial business plan is used to guide the identification and study process by providing the initial outline of the preferred business model and a summary of the public-sector needs, issues, problems, and objectives to be addressed. It is, of course, a legitimate outcome from this work to decide that the public sector has a stand-alone role in the development and supply of advanced traveler information services in the region and move on to detailed business planning with no private-sector collaboration.

From a private-sector perspective, the market study provides the basic information on the potential market size, customer characterization and needs data, and information on potential competitors. This information is used to conduct a feasibility analysis to determine if the market is large enough and presents enough suitable opportunities to warrant further investigation or expenditure. It is also at this point in the business-planning process that the private-sector enterprise would identify the need for partnerships and alliances within both the public and private sectors.

9.4.3 Reference Proposed Activities to the Regional ITS and Transportation Context

We would see this as a public-sector activity, designed to identify and explore possible synergies and collaboration possibilities with other public-sector agencies and initiatives. If a regional ITS architecture has already been developed, then this would be the primary source of information on the possibilities for sensor, data, and information sharing across transportation and ITS initiatives and existing systems. As we discussed in Chapter 7, there are usually a large number of potential opportunities for sharing and collaboration between transportation agencies in a region, due to the cross-cutting nature of ITS and the ease with which information and communication technologies can support shared operation.

9.4.4   Detailed Business Planning and Modeling

This would be part of the collaborative business-planning effort conducted after the initial independent work by both public and private sectors. The proposed business model would be enhanced and detailed in the course of the development of an agreed business plan for the development, deployment, and operation of advanced traveler information products and services. Going beyond the definition of who does what in each step of the traveler information supply chain, the business plan would also identify and define the proposed collaboration and procurement mechanisms to be adopted in support of the respective roles and responsibilities. Making use of the business modeling and planning techniques defined and described in Chapter 6, we suggest that the public and private sectors agree upon and build a shared vision of the proposed ATIS in terms of services, benefits, and values to be provided to the user, as well as the business activities and investment planned to achieve this. At this point in the business-planning process, the specific technologies, products, and services to be developed, acquired, and deployed would be selected based on a thorough evaluation of those currently available in the market, matched to the needs of the system to be deployed.

9.4.5   Advanced Traveler Information Supply-Chain Management

Beyond the business-planning and system deployment stages of the development process, this activity addresses the ongoing management of the overall advanced traveler information supply chain. The activities envisioned include the overall management and monitoring of the seven steps in the supply chain to ensure that the various public- and private-sector activities within each of the steps are properly synchronized, supporting the delivery of the full set of public and private benefits and impacts as planned. This could be either a public- or private-sector activity, or even conducted on a joint basis.

9.5   Conclusion

If we were asked to sum up in a few sentences what we think the overall objective for traveler information systems should be, we would say that the ultimate goal should be readily accessible decision-quality information at all stages of the journey in the best format, at the right time, at the right price, and by the preferred delivery mechanism, in full support of better traveler decision making. Obviously this is a lot easier to say than to actually do, but we hope that we have provided you with a number of useful tools and approaches in the course of reading this book. Starting with a simple and clear mission statement that captures the spirit of what you want to achieve, it is possible to maximize your chances of success as you launch into the actions and investment programs required to make it all happen. Through the application of the structured, collaborative business-planning and modeling approaches we have described, it should be possible to define just the right combination of public and private resources and activities required to generate a win-win situation in your region. While the subject is complicated by the multidisciplinary nature of the qualifications, skills, and experience required, by the interrelationship of the many actions and activities, and by the multiple markets and technology domains involved, there are clear signs that the right approach coupled with appropriate investment levels can create success.

The potential dividend for success is high: the attainment of transportation policy objectives that maximize the effective use of public funds, while maximizing the potential private-sector leverage. Improving the efficiency and effectiveness of the management and operation of our existing transportation-network assets, while improving the user experience, provides us with wonderful justification for past and future investments in our transportation infrastructure.

We hope that in making the investment of your precious time and attention required to read some or all of this book that you find it useful. We have enjoyed the experience of structuring our thoughts and experiences as we developed the materials for the book and have found that our perspectives, thoughts, and ideas regarding ATIS have been irreversibly changed in the process. We hope that we have instilled such changes in your thinking and that, even in a small way, this book has affected and altered your perspective on advanced traveler information, ITS, and transportation.

9.6 Summary

In this chapter we have attempted to build on the definition of the traveler information supply chain by illustrating how it can be used to explore and formulate business models for ATIS. We have introduced you to the concept of the business model and explained what we mean when we use the term. To bring them to life, we have identified and described a series of high-level basic ATIS business models and some more detailed enhanced business models, indicating relationships as well as link activity. These have been related to the practical examples of implementation as introduced in Chapter 2.

We also defined and explained typical business-planning approaches making use of traveler information supply-chain analysis techniques and formal business-planning methodologies. We believe these will serve as a good starting point for public- and private-sector entities embarking on the development and implementation of ATIS. We explored a number of downstream business models for the private sector to consider when developing business models and plans.

This led to the definition of common ground between the sectors and an exploration of one of the key parameters: the value of traveler information. The value of traveler information can be measured in multiple dimensions including the intrinsic and perceived values to the traveler, value to the community, and value to the operating agencies acting on behalf of the community. While some work has been conducted in this area, we identify a need for an agreed approach to the definition and measurement of the value of traveler information. We have described a theoretical approach that we believe holds promise as a potential technique.

The common ground aspect of traveler information systems is particularly interesting to us, as it may well be the key to success in both public and private terms. The current situation is rather complex, with multiple parallel activities involving both public and private sectors at various points in the supply chain within the context of traveler information implementations worldwide. It is obvious that multiple business models can and do operate in parallel in a single region, no matter what model the system under development adopts. This leads to what some observers have described as chaos in the common ground that can be both embraced and leveraged or managed and constrained in order to avoid or mitigate risk. Given that there are very few examples of private-sector commercial success and a general impression that the public sector is not that happy about progress to date, there may be a case for revising our current approach to incorporate a more structured approach to the definition of business models as well as technical arrangements for ATIS.

We believe that it is possible to define an approach to the development and implementation of traveler information systems that maximizes the chances of success for each side of the public-private equation. Through careful planning, requirements definition, deliberate and coherent action, and the application of a combination of direct control and influence or advocacy, it should be possible to create and support a virtuous circle of initial action, complementary action, and reinforcement, leading to sustained progress towards the desired win-win situation. We would like to credit Roger Allport of Halcrow Group in London for defining the concept of the virtuous circle with respect to road user charging. One example of the virtuous circle is illustrated in Figure 9.3.

Here the public sector takes the lead and develops and deploys the required data-infrastructure, data-collection, and data-fusion facilities, providing the fused data to the private sector in return for an agreed revenue stream. The private sector conducts value addition, marketing, and delivery of a set of initial products and services based on the level of quality and quantity of the data. As the market develops, the public sector invests the return revenue stream and other public funds as appropriate to improve and streamline data collection and fusion for both traveler information and traffic-management applications. Such investments improve the quality and quantity of the information stream to the private sector, enabling them to improve the quality and sophistication of the products and services being delivered to the users. This in turn increases the potential revenue achievable and increases the size of the return revenue stream to the public sector, which in turn enables the next incremental improvement in the data infrastructure, collection, and fusion capabilities. This cycle repeats until the desired endstate is reached and the best quality traveler information is available over the widest coverage area to the largest number of users, in the forms they wish to use.

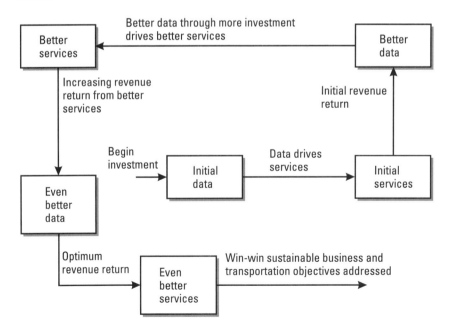

Figure 9.3 The virtuous circle for traveler information.

The adoption of this type of approach would address the issue of what has been called the tipping point, as defined in [2]. The theory is that new products, services, or ideas when introduced have to reach some critical mass or tipping point at which their use or adoption grows at an extremely accelerated rate. If traveler information products and services obey this theory, then it would be a tremendous investment to support them heavily in the early stages in order to receive payback once the tipping point is achieved. Historical information about the introduction and growth of similar products and services, such as the Internet, cellular telephones, TV, and radio indicate that this theory may well be relevant in this context.

Obviously, the adoption of this approach would require the establishment, management, and maintenance of a coordinated approach between the public and private sectors. Public-sector objectives and policies would have to be aligned to support the initial investment and the continued investment in data infrastructure collection and fusion. Private-sector business plans would have to be aligned to provide continuous service and product improvement.

This is not the only possible virtuous circle. We are convinced that with the right approach to planning and development, you can identify and apply not just the right model for day one, but the evolutionary path that will take you in a sustainable manner to the desired goals. The main thing is to be extremely cautious with the use of investment funds and think twice before the adoption of a single narrow path to the future.

The common ground also extends to the shared funding of the various activities in the traveler information supply chain. As money is a constrained resource and the market for traveler information services seems to be at an embryonic stage of development, the careful selection of funding mechanisms is vital. Aside from the "public starts, private joins in" approach, there are additional options for the public sector to identify and utilize opportunities for market development and business development support in recognition of the opportunity to satisfy public objectives in the course of private business enterprise.

Looking again at the downstream business models described earlier in this chapter, a public role could be defined as either a shadow sponsor or anchor client for the products and services to be delivered. For example, in the advertising model, the public sector could act as a real or shadow sponsor, either buying advertising space for public service messages or simply subsidizing the private sector on a per-case or performance-related basis through shadow payments. Many of the other downstream models could be adapted to this use, enabling the private sector to have the freedom to develop and deliver the services, yet have access to public support in recognition of the effects on traveler behavior. Indeed, the TravInfo® operation in the San Francisco Bay Area seems to be moving in this direction with the recent adoption of performance-related fees for operating the system.

When we considered the VICS business model in Japan and the Trafficmaster™ example in the United Kingdom, several interesting points were raised. Both of these examples make use of a national rather than regional business model for ATIS, and both seem to be reasonably successful. Should future business models for ATIS be regional in nature and coverage, or national? Should there be a federal or central government subsidy driving national-level private-sector involvement instead of a regional or metropolitan-area-specific patchwork?

These questions stimulate interest in examining a wider range of options and relate to the cultural differences between the public and private sectors discussed in Chapter 4. While the public sector has a project, corridor, or regional focus, the private sector demonstrates a desire to address the subject of traveler information on a national basis at a minimum. Perhaps this is because one of the largest potential market segments, in-vehicle information services, is driven by national and international interests. Perhaps there is a role to be defined for a national transportation data wholesaler offering comprehensive transportation-management and decision-quality traveler-information-related data. Questions like these have no universal answer, but we hope that we have shown you over the course of this book that techniques are available to help you to define the best solution for your circumstances.

We hope that you have found the investment of your time and attention in reading this book to be a value proposition and that the information and knowledge we bring will be of value to you. If we have influenced your thinking with respect to ATIS in even a small way, then we will have succeeded in our objective.

References

[1]   Clemen, R. T., *Making Hard Decisions: An Introduction to Decision Analysis*, 2nd ed., Belmont, CA: Duxbury Press, 1996.

[2]   Gladwell, M., *The Tipping Point: How Little Things Can Make a Big Difference*, Boston, MA: Little Brown Book Publishers, 2000.

About the Authors

Bob McQueen is an internationally recognized expert in the application of information and telecommunications technologies to transportation systems management and operations. With a background in civil, transportation, and traffic engineering, he has developed a world-class reputation as an innovative business and technical specialist in intelligent transportation systems (ITS) topics. He has a bachelor's degree in civil engineering from the University of Strathclyde, Glasgow, Scotland,  and a master's degree in highways and transportation from the City University, London, England.

Mr. McQueen's experience includes high-level consulting assignments for national and local government agencies in the United States, Europe, the Middle East, and Southeast Asia, spanning a range of ITS activities from policy development, requirements analysis, conceptual design, and system architectures. His recent activities focus on the development of education and professional development programs for ITS and the definition and development of organizational structures and business models for management and operations of advanced technology applications. He is a vice president and division manager of ITS Services for PBS&J, based in Orlando, Florida.

Rick Schuman is one of the world's leading ITS consultants and a national leader in the development and application of traveler information services and technologies. He has managed several consulting projects in the areas of traveler information and ITS data-collection planning and implementation for several clients, including the U.S. Department of Transportation (DOT), ITS America, the American Association of State Highway and Transportation Officials (AASHTO), and the Florida, Texas,  and Arizona DOTs. His current emphasis is on innovative data collection, ITS business models, and 511 service definition and delivery.

Mr. Schuman has a bachelor's degree in electrical engineering from Boston University, Boston, Massachusetts, and a master's degree in systems engineering from George Mason University, Fairfax, Virginia. He is manager of traveler information systems for PBS&J, based in Orlando, Florida.

Kan Chen is professor emeritus of electrical engineering and computer science at the University of Michigan, Ann Arbor, Michigan. While in Michigan, Dr. Chen founded the ITS research and educational program in the 1980s and helped with the development of the national ITS program for the United States. He also directed the interdisciplinary Ph.D. program in urban, technological, and environmental planning at the University of Michigan for 6 years and taught courses in technology assessment and planning.

Dr. Chen received a bachelor's degree from Cornell University, Ithaca, New York, and a Ph.D. from the Massachusetts Institute of Technology, Cambridge, Massachusetts, both in electrical engineering. He is a fellow of the Institute of Electrical and Electronic Engineers and the American Association for the Advancement of Science. He has written and edited eight books, the most recent being the *ITS Handbook 2000* (Artech House, 1999).

As an ITS consultant, Dr. Chen has served as the Technical Review Team chairman for the U.S. National Architecture Program. He is currently one of the two U.S. coordinators for the Atlantic Project, within which he is also leading the effort to benchmark ATIS activities between Europe and North America. Dr. Chen is based in San Francisco, California.

Index

Recent Titles in the Artech House ITS Library

John Walker, Series Editor

Advanced Traveler Information Systems, Bob McQueen, Rick Schuman, and Kan Chen

Advances in Mobile Information Systems, John Walker, editor

Incident Management in Intelligent Transportation Systems, Kaan Ozbay and Pushkin Kachroo

Intelligent Transportation Systems Architectures, Bob McQueen and Judy McQueen

Introduction to Transportation Systems, Joseph Sussman

ITS Handbook 2000: Recommendations from the World Road Association (PIARC), PIARC Committee on Intelligent Transport (Edited by Kan Chen and John C. Miles)

Positioning Systems in Intelligent Transportation Systems, Chris Drane and Chris Rizos

Sensor Technologies and Data Requirements for ITS, Lawrence A. Klein

Smart Highways, Smart Cars, Richard Whelan

Tomorrow's Transportation: Changing Cities, Economies, and Lives, William L. Garrison and Jerry D. Ward

Vehicle Location and Navigation Systems, Yilin Zhao

Wireless Communications for Intelligent Transportation Systems, Scott D. Elliott and Daniel J. Dailey

For further information on these and other Artech House titles, including previously considered out-of-print books now available through our In-Print-Forever® (IPF®) program, contact:

Artech House
685 Canton Street
Norwood, MA 02062
Phone: 781-769-9750
Fax: 781-769-6334
e-mail: artech@artechhouse.com

Artech House
46 Gillingham Street
London SW1V 1AH UK
Phone: +44 (0)20 7596-8750
Fax: +44 (0)20 7630 0166
e-mail: artech-uk@artechhouse.com

Find us on the World Wide Web at:
www.artechhouse.com